MATERIALS SCIENCE AND

NEUTRON DIFFRACTION

PRINCIPLES, INSTRUMENTATION AND APPLICATIONS

MATERIALS SCIENCE AND TECHNOLOGIES

Additional books in this series can be found on Nova's website under the Series tab.

Additional e-books in this series can be found on Nova's website under the e-book tab.

MATERIALS SCIENCE AND TECHNOLOGIES

NEUTRON DIFFRACTION

PRINCIPLES, INSTRUMENTATION AND APPLICATIONS

XINZHE JIN
EDITOR

NOVINKA

New York

Library of Congress Cataloging-in-Publication Data

Neutron diffraction : principles, instrumentation and applications / editor: Xinzhe Jin (Cryogenic Science Center, High Energy Accelerator Research Organization (KEK), Ibaraki, Japan).
pages cm
Includes bibliographical references and index.
ISBN 978-1-62808-725-3 (soft cover)
1. Neutrons--Diffraction. I. Jin, Xinzhe.
QC793.5.N4628N484 2013
539.7'213--dc23

2013029197

Published by Nova Science Publishers, Inc. † New York

CONTENTS

Preface		**vii**
Chapter 1	Principles and Instrumentation of Neutron Diffraction Measurement for Material Science *Xinzhe Jin*	**1**
Chapter 2	Neutron Diffraction Study of Isotopic Effect in Arrangement of Hydrogen Atoms in Solid Solutions of the Ti-N-H, Ti-N-D and Ti-N-H-D Systems *I. Khidirov*	**13**
Chapter 3	Study on Residual Strain Induced by Thermal Hysteresis in a Composite Material *Xinzhe Jin*	**37**
Chapter 4	The Neutron Diffractometers for Fatigue Study of Engineering Metals *E-Wen Huang, Soo Yeol Lee and Chung-Hao Chen*	**47**
Index		**75**

PREFACE

The study of neutron diffraction has evolved significantly in recent years as new materials technology has been researched and developed. On the other hand, there are very few books about the introduction of neutron diffraction. In this book, I have introduced the recent research information in a format that is easy to understand, to make the technology more accessible.

Xinzhe Jin
Apr. 2013

In: Neutron Diffraction
Editor: Xinzhe Jin

ISBN: 978-1-62808-725-3

Chapter 1

PRINCIPLES AND INSTRUMENTATION OF NEUTRON DIFFRACTION MEASUREMENT FOR MATERIAL SCIENCE

Xinzhe Jin*

High Energy Accelerator Research Organization (KEK),
Ibaraki, Japan

ABSTRACT

The neutron beam has been applied extensively in scientific studies and in industrial fields. Studies on neutron diffraction have been promoted for many materials, especially composite materials and light-element materials. In this paper, the measurement principles and methods of some typical neutron diffractometers are presented. Also, the multi-peak analysis method for the diffraction pattern of composite materials is described from a statistical physics viewpoint.

Keyword: Neutron diffraction, multi-peak analysis, composite material

* Corresponding author: shinte@post.kek.jp; kinnsinntetu@yahoo.co.jp.

INTRODUCTION

Neutron diffraction uses beam phenomena such as scattering, reflecting, diffraction, and interference. Neutron diffraction is usually indicating elastic scattering that is carried out for nondestructive identification.

For elastic scattering, the beam energy is typically less than 100 eV, and the elastic scattering region in the beam energy depends on the material. Neutron diffraction is also subject to Bragg's law for X-ray diffraction.

When a substance having a periodic structure such as crystal is irradiated with a beam having some penetration depth, a strong emitted beam can be observed in the particular angles or in the particular beam energies under the condition, because the interference affects the reflecting.

The intense beam by interference is undetectable in other angles or beam energies, because overlap of beam wave does not occur, despite the fact that it is reflecting.

Neutron diffraction using slow neutrons has the property that the slow neutrons are absorbed by elements such as Li, B, Cd, and Gd [1]. Therefore, experiments should verify beforehand whether these elements are contained in the measurement sample. Low-intensity patterns are detected by neutron diffraction measurement, if the material contains a large amount of the abovementioned elements.

However, some cases utilize this characteristic to obtain a background-cleaned diffraction pattern by coating the sample holder, except the beam path, by using these elements. Neutron diffraction patterns can still be obtained clearly in compounds with a small amount of the above elements, as in the neodymium magnet $Nd_2Fe_{14}B$.

The neutron beam has a larger penetration depth than an electronic beam and an X-ray with similar energy.

This property is very useful in research on composite materials to obtain internal information of single materials. Neutron diffraction can observe light elements such as H, C, N, and O, which are undetectable in X-ray diffraction. Therefore, neutron diffraction has been applied to studies on hydrogen storage alloys, oxygen atomic coordinates in high-temperature superconductivity, and organic materials.

In addition, neutron diffraction is used for analysis of the magnetic structure of magnetic materials, although this paper mentions the lattice structure analysis without a magnetic structure analysis.

MEASUREMENT PRINCIPLES

Neutron diffraction measurement has four major components: beam source, sample, detector, and measurement environment. The beam source usually is fixed. Measurement methods for neutron diffraction can be categorized into angular variance and wavelength dispersion types. The angular variance type uses a single wavelength, and the diffraction patterns obtained depend on the diffraction angle. The wavelength dispersion type measures the multi-wavelength beam time of flight (TOF) from a mechanical chopper based on the de Broglie wave rule.

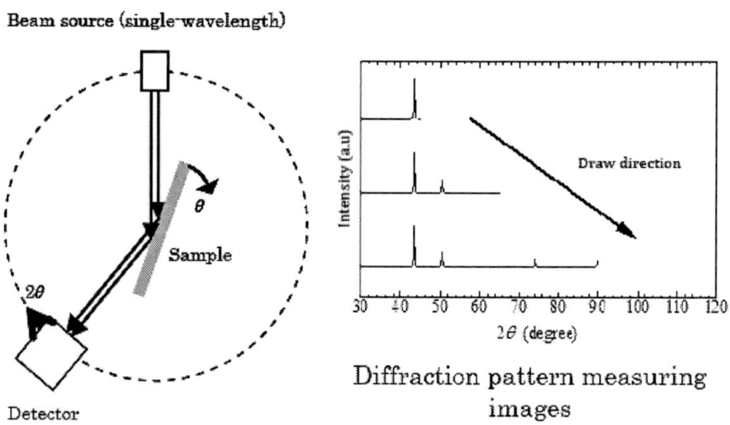

Figure 1. Neutron or X-ray (Cu-Kα_1) diffraction measurement diagram illustrating angular dispersive method by turning of sample and detector.

In the angular variance type of neutron diffraction, measurements are carried out with a moving sample or with an *in situ* method. The measurement principle of the moving sample method, shown in Figure 1, is usually adopted in X-ray diffraction (Cu Kα) in a general experimental laboratory. The *in situ* method, shown in Figure 2, uses multi-detectors to measure the angle dependence of the beam diffraction.

In wavelength dispersive neutron diffraction, a Takumi neutron diffractometer was recently provided at MLF/J-PARC for studying the TOF method on engineering materials [2, 3]. Multiple peaks of neutron diffraction can be obtained simultaneously by using two fixed detectors, as shown in Figure 3. For example, in a wire sample, diffraction patterns in the longitudinal and radial directions of a sample can be obtained simultaneously by placing the sample at an angle of 45° from the incident beam line. The neutron diffraction measurement can be performed by loading the sample from room temperature to high temperature, or from room temperature to cryogenic temperature, by using the load frame.

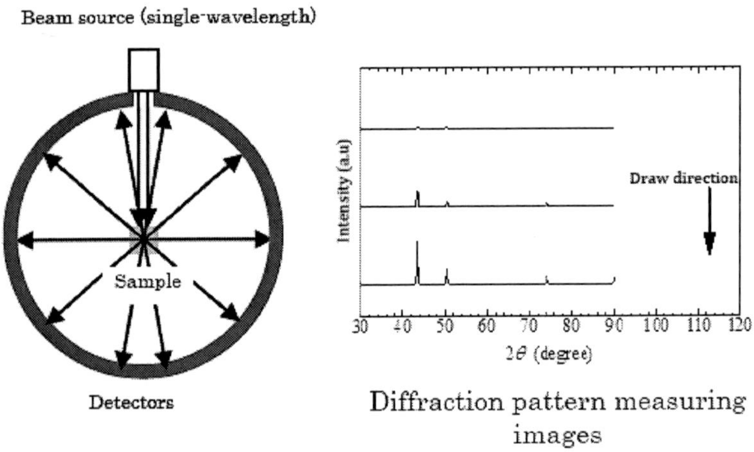

Figure 2. *In situ* beam diffraction measurement diagram.

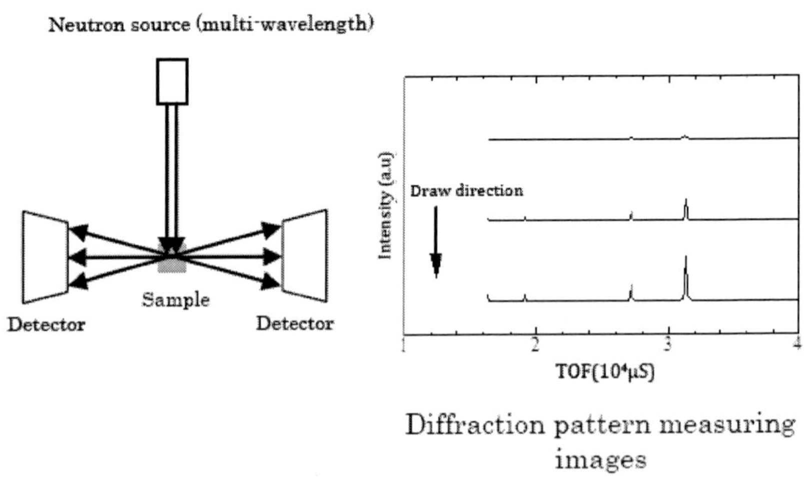

Figure 3. Neutron diffraction measurement diagram of TOF method by the J-PARC Takumi diffractometer.

Calibration of the TOF and lattice spacing *d* is usually performed by using standard sample such as CeO_2 powder. The neutron incident beam size has two modes, high intensity and high resolution, with a sample slit width of 5 and 2 mm, respectively. Takumi's design corresponds to 1 MW proton beam power. Recently, experiments were conducted at a proton beam power of 300 kW.

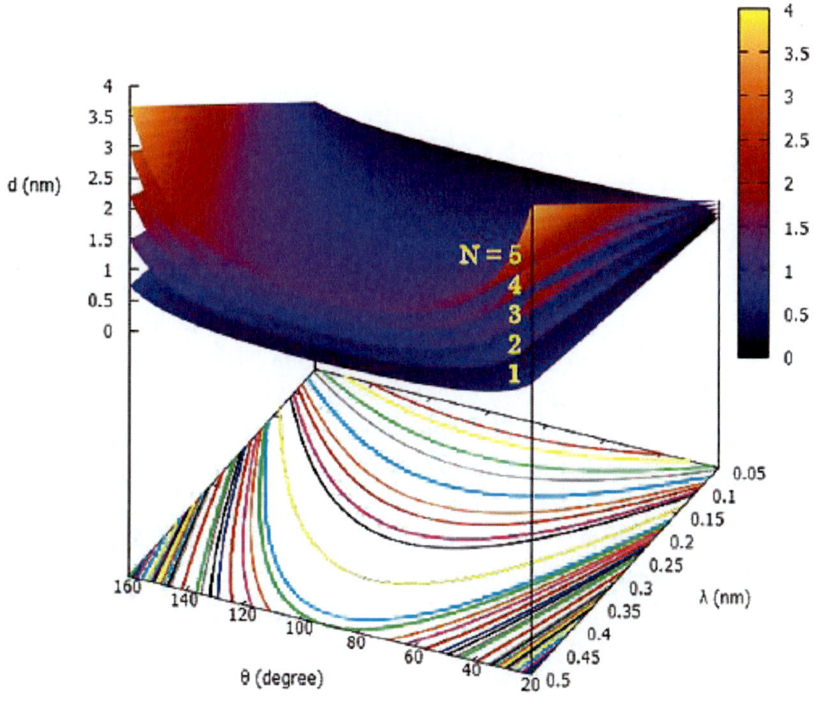

Figure 4. 3D plot of Bragg's condition for neutron diffraction.

The quantum beam diffraction according to Bragg's condition is as follows:

$$2d\sin\theta = n\lambda, \tag{1}$$

where *d* is the lattice spacing, θ is the angle between the incident beam and the lattice plane, *n* is a positive integer, and λ is the wavelength. As an example, the 3D plot of the function for n = 1, 2, 3, 4, 5 is shown in Figure 4.

In TOF neutron diffraction by de Broglie wave, λ is

$$\lambda = \frac{h}{mv} = \frac{ht}{ms}. \tag{2}$$

It is known the Planck's constant is h and the neutron mass is m. Therefore, TOF t can be obtained with flight distance s as

$$t = \frac{2msd\sin\theta}{h} = f(s,d,\theta). \tag{3}$$

For example, the diagram of diffraction patterns with the function $t = f(d, \theta)$ is shown in Figure 5. Considering the default value t_0, t is

$$t = f(s, d, \theta) + t_0. \tag{4}$$

The lattice spacing d having scattering intensity in the diffraction pattern takes discrete values by the scattering synthesis of all atoms in the lattice. Some background of crystallography is presented in the following.

Given the equal distance of parallel lattice planes, the lattice spacing d is related to the lattice structure in a lattice plane by the method of Miller indices. For example, some Miller indices and the corresponding lattice planes in a unit cell are shown in Figure 6.

In neutron diffraction in the (001) plane, as shown in Figure 7, the optical path difference between two adjacent lattice planes is $n\lambda$. Since the lattice spacing for the (002) plane is half of that for the (001) plane, the optical path difference is $2n\lambda$.

If the same atoms are located on the center plane along the c axis in the unit cell, the number of neutron interfering planes for (002) is larger than that for (001).

For example, since the atoms in Cu of a face centered cubic (FCC) structure are located at the center of each surface, no peak of the (001) plane can be observed in the neutron diffraction pattern. This result is the incoherent effect between closed planes due to the difference of the half wavelength, shown in Figure 7 by the dashed lines with arrows. Therefore, this effect has an influence on the intensity of the beam diffraction, such that (00l) planes have a different surface regularity depending on whether l is an even or odd number.

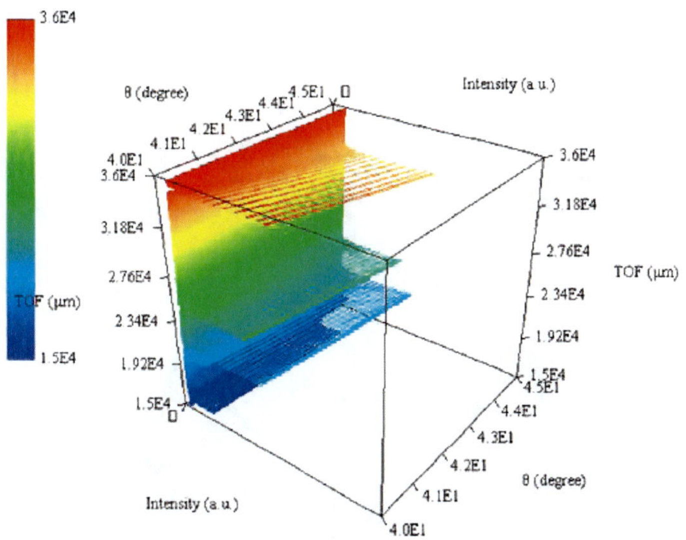

Figure 5. Neutron diffraction simulation for Cu by TOF method with multi-detectors around sample.

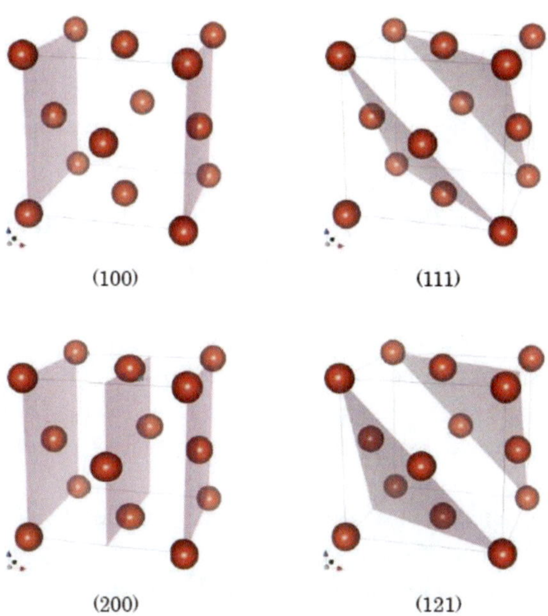

Figure 6. Lattice planes for Miller indices of 100, 111, 200, and 121.

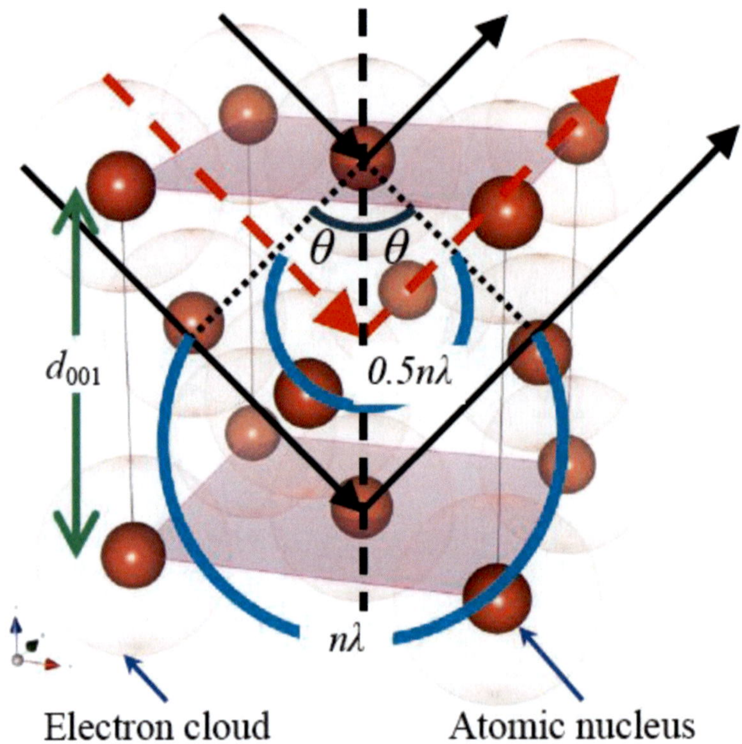

Figure 7. Neutron diffraction diagram in unit cell for (00*l*) lattice planes.

Multi-Peak Analysis of Neutron Diffraction Pattern

It is known that information of the lattice structure, such as lattice parameters and site occupancy, can be obtained by using the Rietveld analysis for a powder sample [4, 5]. For a composite sample having residual strain and also having texture, we proposed a multi-peak analysis to obtain the lattice parameter [6-9].

By using the lattice constant of the model obtained by the least squares method with a weighted average in the multi-peak analysis, the obtained residual strain is close to the macro strain value. In the multi-peak analysis of the composite sample having residual stress, the difference of the diffraction elastic coefficients in the lattice planes should be considered.

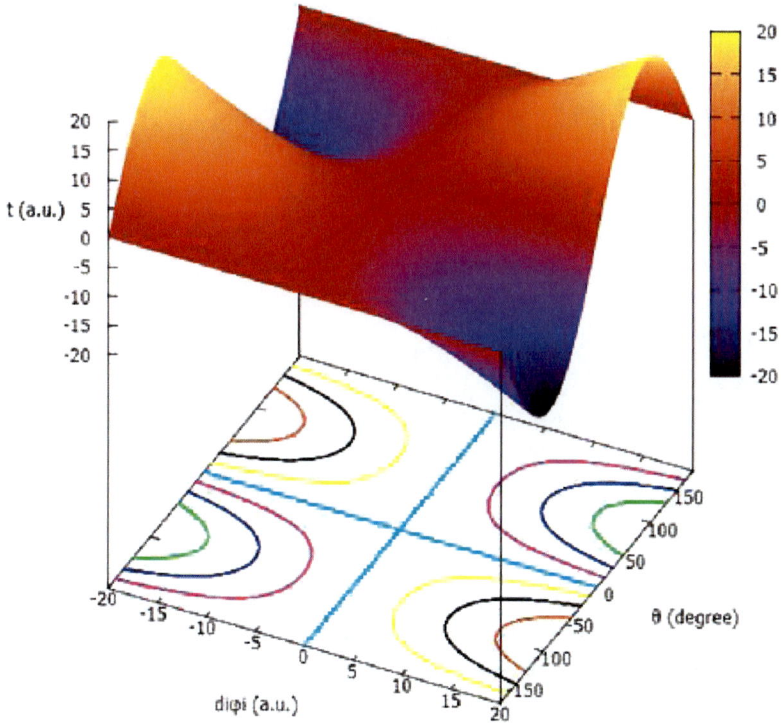

Figure 8. 3D plot of neutron diffraction condition for multi-peak analysis.

With the diffraction elastic coefficient φ_i, the neutron diffraction condition of the TOF type is

$$t = f(s, d_i\varphi_i, \theta) + t_0 = \frac{2msd_i\varphi_i \sin\theta}{h} + t_0 . \tag{5}$$

The 3D plot of equation (5) is shown in Figure 8. In the equation, d_i has dependence on the lattice structure of a material with discrete values. The least squares method for a cubic crystal is

$$t_i = Dax_i\varphi_i\sin\theta + t_{0,} \tag{6}$$

$$S = (t_i - (Dax_i\varphi_i\sin\theta + t_0))^2, \tag{7}$$

$$\frac{\partial S}{\partial a} = 0, \tag{8}$$

$$aD\sin\theta\, \Sigma(x_i^2 \varphi_i^2) + t_0 \Sigma(x_i \varphi_i) - \Sigma(t_i x_i \varphi_i) = 0, \tag{9}$$

$$a = (\Sigma(x_i \varphi_i\, (t_i - t_0)))/(D\sin\theta\, \Sigma(x_i^2 \varphi_i^2)). \tag{10}$$

In the multi-peak analysis, the statistical value $\hat{\varphi}$ is

$$\hat{\varphi} = \frac{\sum (x_i^2 \varphi_i^2) \sum (x_i (t_i - t_0))}{\sum x_i^2 \sum (x_i \varphi_i (t_i - t_0))} \tag{11}$$

CONCLUSION

The measurement principles and methods of some typical neutron diffractometers are presented. Also, the multi-peak analysis method for the composite materials is described from a statistical physics viewpoint with calculation. That statistical method with the multi-peak analysis in neutron diffraction pattern will be applied more in engineering and academic fields.

REFERENCES

[1] Dunning JR, Pegram GB, Fink GA and Mitchell DP, 1935, *Physical Review* 48, 265.

[2] Harjo S, Ito T, Aizawa K, Arima H, Abe J, Moriai A, Iwahashi T and Kamiyama T, 2011, *Mater. Sci.* Forum 681, 443.

[3] Harjo S, Aizawa K, Ito T, Arima H, Abe J, Moriai A, Sakasai K, Nakamura T, Nakatani T, Iwahashi T and Kamiyama T, 2010, *Mater. Sci. Forum* 652, 99.

[4] Izumi F and Ikeda T, 2000, *Mater. Sci. Forum* 321–324, 19.

[5] Torii S and Moriai A, 2006, *Physica B* 385, 1287.

[6] Jin X, Nakamoto T, Ito T, Harjo S, Kikuchi A, Takeuchi T, Tsuchiya K, Hemmi T, Ogitsu T and Yamamoto A, 2012, *Supercond. Sci. Technol*. 25, 065021.

[7] Jin X, Nakamoto T, Harjo S, Ito T, Ogitsu T, Tsuchiya K, Yamamoto A, Kikuchi A, Takeuchi T and Hemmi T, 2012, J. *Alloys Compds.* 535, 124-128.

[8] Jin X, Oguro H, Nakamoto T, Awaji S, Yamamoto A, Tsuchiya K, Ogitsu T, Kikuchi A, Takeuchi T, Harjo S, and Ito T, 2012, *Cryogenics* 52, 805-809.

[9] Jin X, Nakamoto T, Tsuchiya K, Ogitsu T, Yamamoto A, Kikuchi A, Takeuchi T, Harjo S, Ito T, and Tomota Y, 2013, *IEEE Trans. Appl. Supercond.* 23, 6000704.

In: Neutron Diffraction
Editor: Xinzhe Jin
ISBN: 978-1-62808-725-3

Chapter 2

NEUTRON DIFFRACTION STUDY OF ISOTOPIC EFFECT IN ARRANGEMENT OF HYDROGEN ATOMS IN SOLID SOLUTIONS OF THE Ti-N-H, Ti-N-D AND Ti-N-H-D SYSTEMS

I. Khidirov*

Institute of Nuclear Physics, Uzbekistan Academy of Sciences, Tashkent, Uzbekistan

ABSTRACT

It was shown that isotopic different the $TiN_{0.26}H_{0.15}$, $TiN_{0.26}D_{0.15}$ solid solutions in a high-temperature disordered state (the space group $P6_3/mmc$) have identical structural characteristics. However, as temperature decreases an isotopic effect appears. This effect means that unlike hydrogen solid solution, in the deuterium solid solution the ordering starts at much higher temperature; the ordered monoclinic phase has a larger homogeneous region; the interstitial ^2D and ^1H distribution over interstices have different characters and the blocking effect between interstitial atoms is more pronounced. It is interesting that the displacement of the hydrogen atoms from the center of the tetrahedron along the *c*-axis has different signs in the disordered and ordered solid

* E-mail: khidirov@inp.uz.

solutions. The crystal structure (space group $P\bar{3}m1$) of the solid solution with the combined isotope composition $TiN_{0.40}H_{0.19}D_{0.19}$ was studied by the neutron structure analysis. The isotopic effect was found in distribution of the hydrogen isotopes H and D: the isotopes are located over the same type of tetrahedral interstices but have different coordinates z, i.e. the splitting of a set of the tetrahedral interstices 2d into the hydrogen (H) and the deuterium (D) was observed. It follows that in the "layered" hexagonal ordered structure of the $TiN_{0.40}H_{0.19}D_{0.19}$ solid solution, two different isotopes of the hydrogen can not be on one plane perpendicular to the threefold axis. Crystal-chemistry analysis of environment of the H and the D was carried out and the essential differences in coordination polyhedrons surrounding the two isotopes were revealed. The analysis of these results and also literature data shows that in a number of metal - hydrogen systems the hydrogen and the deuterium should be considered as independent chemical components.

Keywords: Solid solutions, $TiN_{0.26}H_{0.15}$, $TiN_{0,40}H_{0,21}$ $TiN_{0.26}D_{0.15}$, $TiN_{0,26}H_{0,075}D_{0.075}$, $TiN_{0,40}H_{0,19}D_{0,19}$, crystal structure, hydrogen, deuterium, interstices, isotopic effect, neutron diffraction

INTRODUCTION

Due to such properties as high melting temperature, small specific gravity, high solidity, corrosion stability in many aggressive medium, the titanium is an universal and valuable material for many constructions in various branches of new techniques [1, 2]. Titanium properties essentially depend on the quantity and character of distribution of the impurity of nonmetallic elements introduced into them, in particular, such as carbon, nitrogen, oxygen and hydrogen. Hydrogen is one of the most harmful and frequently occurring impurities in titanium. In some cases harmful influence of hydrogen is neutralized by connecting it in more stable phases than simple titanium hydrides: hexagonal nitrohydrides, carbohydrides and others. These complex phases are more stable than double titanium hydrides [2]. Moreover, hexagonal titanium nitrohydrides and carbohydrides are the solid solutions of nitrogen (or carbon) and hydrogen in the α-Ti lattice with the high content of hydrogen. So $TiN(C)_xH_y$ solid solution because of the high thermal stability ($T \geq 1270$ K) and corrosion stability can be used as the hydrogen accumulator for hydrogen energetic in extreme conditions: at rather high temperatures and in aggressive environments. For purposeful obtaining of interstitial solid solution of nitrogen and hydrogen in the α-Ti lattice with the given service characteristics it is necessary to know the structure features of this material.

STATE OF THE QUESTION

Several works are devoted to structure investigations of the TiN_xH_y interstitial solid solution with different metalloid concentration over the wide temperature intervals [3-12]. In the Ref. [4] the hexagonal type of the unit cell of titanium nitrohydride was pointed. However, the question about localization of the hydrogen atoms was opened. In [5] the TiN_xH_y samples were been investigated and the conclusion was made that hydrogen atoms are located in the octahedral interstices. The work [6] has clarified the question about the crystal structure of hexagonal titanium nitrohydrides. In this work the regular neutron diffraction study of these compounds was carried out in a wide homogeneity ranges, and the conclusion was that these compounds undergo the phase transition of order - disorder type on both the nitrogen (carbon) and the hydrogen sublattice ($P6_3/mmc$ (α-phase) → $P\bar{3}m1$ (α′-phase)). The nitrogen (carbon) atoms occupy only octahedral interstices but the hydrogen atoms can occupy both octahedral and tetrahedral interstices. Concrete distribution of interstitial atoms depends on the composition and the temperature. In the same work, the conclusion was that there is a correlation between the distribution of the nitrogen (carbon) and the hydrogen atoms: the hydrogen sublattice is not ordered if the nitrogen (carbon) sublattice is disordered. However, some questions are unsolved in this work too. In particular, in the strongly defective solutions, such as the $TiN_{0.40}H_{0.21}$, $TiN_{0.32}H_{0.21}$ and the $TiN_{0.26}H_{0.15}$, the full ordering of nitrogen atoms was not observed. At the same time, according to "the diagram of branching", it is possible to expect either the full ordering or the decay of these solid solutions. The coordinates of the tetrahedron centers were taken as the free parameters of the hydrogen atoms without confirmation by calculations. In [8] it is shown that after the annealing at temperatures 800 -700 °C the ordered α′-phase (the space group $P\bar{3}m1$) passes into the γ-phase within the nitrogen concentration interval of x = 0.32÷0.40. The crystal structure of this phase is monoclinic (pseudo-rhombic) with the unit cell parameters: a = 5.163(3), b = 3.012(2), c = 4.828(1) Å, β = 89.82(7)°, the space group is C2/m. On the basis of the neutron diffraction analysis by the Rietveld method the coordinates of all basic atoms (including hydrogen) has been determined. In the new monoclinic structure (the structural type of anti-$AuTe_2$) the layer character of the initial hexagonal phases conserves. Later on in [9] the consecutive process of ordering-decay and of antiphase domains ordering was found out at temperatures lower than the decay temperature 600 ^{0}C. Similar investigations of the TiN_xD_y interstitial solid solution were carried out for the $TiN_{0.26}D_{0.15}$ in [11, 12]. In the last work [12] a number of distinctive structural features of the $TiN_{0.26}H_{0.15}$ and $TiN_{0.26}D_{0.15}$ solid solutions caused by their different isotopic compound was found out.

It is of interest to carry out the neutron diffraction studies of both TiN_xD_y and $TiN_xH_{y/2}D_{y/2}$ solid solutions with different concentration in the wide temperatures interval. The obtained results and comparison of them with the results of the study of TiN_xH_y solid solutions can give the valuable information about the role of hydrogen, its isotopes and energy of deformative interaction of interstitial atoms between themselves and solvent atoms in the formation of structure and phase relations of hydrogen-containing solid solutions of the nitrogen atoms in the α-Ti lattice. It should be noted that the deuterium and hyrogen nuclei have quite different amplitudes of neutron scattering (b_D = 0,667·10^{-4} Å and b_H = -0.374·10^{-4}Å) and deuterium has little incoherent background in comparison with the isotope ^{1}H [13]. Therefore, the neutron diffraction study of deuterium solid solution allows determining the position of deuterium and hydrogen atoms in a crystal lattice with a high accuracy. It is of interest to study solid solution with mixed isotopic composition $TiN_xH_{y/2}D_{y/2}$ because it will allow to answer the question how universal is the isotopic separation effect of the hydrogen sublattice. May be, it is similar to the CeD_2H compound [14], in which Somenkov V. A., etc. observed the isotopes separation over the different interstices types.

The aim of this work is a generalization the results of the neutron diffraction studies of the structure features of solid solutions in the Ti-N-D and Ti-N-H-D systems, having different concentration, in the wide temperature interval and a comparative analysis of these results with the results of investigation of the Ti-N-H system solid solutions.

EXPERIMENT

Neutron diffraction patterns of powdery samples were obtained at the neutron diffractometer DN-500 of the INP AS of Uzbekistan (λ = 1.085 Å) [15]. Calculations and refinement of the structure characteristics were carried out by the Rietveld full-profile method [16]. The X-ray diffraction patterns were obtained at the diffractometer DRON-3M (CuK_{α} – radiation, $\lambda = 1.5418$ Å). The samples were prepared by the self-propagating high-temperature synthesis [17] or by the method of Siverts [3]. All samples were followed by 6-h homogenizing anneals in degassed and sealed silica glass ampoules at temperature 1270 K and by air quenches. The elements content has been determined by the chemical analysis.

The neutron and the X-ray diffraction were applied to study of the ordering in the $TiN_{0.26}H_{0.15}$, $TiN_{0.26}D_{0.15}$, $TiN_{0.26}H_{0.075}D_{0.075}$ and $TiN_{0.40}H_{0.19}D_{0.19}$ solid solutions [6, 8, 9, 18, 19]. The X-ray powder diffractions and the neutron patterns show that the $TiN_{0.26}H_{0.15}$ and $TiN_{0.26}D_{0.15}$ solid solutions obtaining by quenching from temperature 1270 K are single–phases: the hexagonal space group is $P6_3/mmc$ (disordered, α-phase), $a = 2.978 \pm 0.002$, $c = 4.795 \pm 0.003$ Å [9] and $a = 2.983$, $c = 4.805$ Å [12], accordingly. The $TiN_{0.26}H_{0.075}D_{0.075}$ solid solution was prepared using equal proportions of the $TiN_{0.26}H_{0.15}$ and $TiN_{0.26}D_{0.15}$ by sintering in evacuated and sealed quartz ampoule; it was sintering under special regime selected by us. The final product was obtained after the annealing at 1270 K (for 6 h) followed by quenching in air. Accorning to the X-ray and neutron diffraction patterns, the obtained sample of the $TiN_{0.26}H_{0.075}D_{0.075}$ solid solution was of the single-phase with the partially disordered hexagonal structure (the space group $P6_3/mmc$) and had the lattice parameters: $a = 2.987 \pm 0.007$, $c = 4.839 \pm 0.003$ Å [18]. According to the X-ray diffraction pattern, the synthesized $TiN_{0.40}H_{0.19}D_{0.19}$ solid solution, obtained by method of Siverts was single-phase and homogeneous with the ordered structure (space group $P\bar{3}m1$; α′-phase) and had the hexagonal (trigonal) unit cell with the lattice parameters $a = 3.009$, $c = 4.875$ Å [19]. Thus both the $TiN_{0.26}H_{0.075}D_{0.075}$ and $TiN_{0.40}H_{0.19}D_{0.19}$ solid solutions have the ordered structure described in the space group $P\bar{3}m1$ (α′-phase).

RESULTS AND DISCUSSION

The processing of the neutron diffraction pattern, (Fig. 1 *a*) of the $TiN_{0.26}H_{0.15}$ solid solution quenched from 1270 K showed that all the nitrogen atoms statistically occupy the octahedral interstices 2(a), hydrogen atoms occupy the octahedral interstices 2(a) (50 %) and thc tetrahedral interstices 4(f) (50 %) in the space group $P6_3/mmc$ (α-phase). A diffuse neutron scattering is observed is observed at the Bragg angle $2\theta = 11\text{-}18^0$ (Fig. 1b) after quenching the sample at temperatures 1170-1070 K. The Long–range ordering on nitrogen atoms corresponding to a structure of the anti-CdI_2 type ($P\bar{3}m1$) becomes noticeable after 12 h at temperatures 920 – 820 K (Fig 1c). The neutron data processing shows that at these temperatures the hydrogen atoms occupy only one type of the tetrahedrons 2 (d) and the nitrogen atoms are partially disordered, in preference occupying the octahedral positions 1 (a) in the space group $P\bar{3}m1$(α'-phase). The long–range ordering on the nitrogen atoms corresponding to a structure of the anti - CdI_2 type ($P\bar{3}m1$) becomes noticeable after 12 h at temperatures 920 – 820 K (Figure 1c).

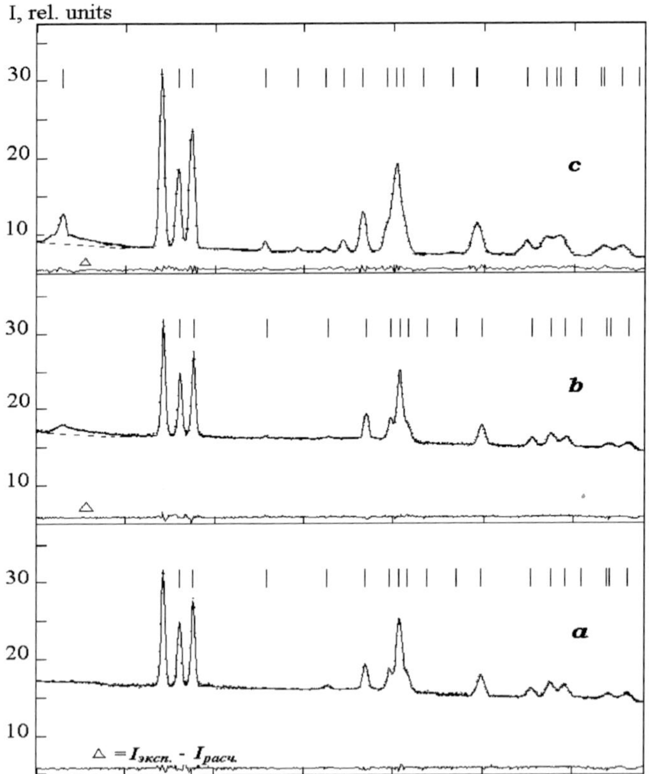

Figure 1. Neutron diffraction patterns of the $TiN_{0.26}H_{0.15}$ quenched from 1170 K (a); 1070 K (b) and 890 K (c) (points show experimental data, solid line is the fitted curve and $\Delta = I_{exp.} - I_{calc}$).

The neutron data processing shows that at these temperatures the hydrogen atoms occupy only one type of the tetrahedrons 2 (d) and the nitrogen atoms are partially disordered, in preference occupying the octahedral positions 1 (a) in the space group $P\bar{3}m1$(α'-phase). The increase of exposure up to 28 h at temperature 870 K causes a complete disintegration of the solid solution into a pair of phases, one having a lower and the other − a higher nitrogen percentage than that of the $TiN_{0.26}H_{0.15}$[9]. One phase is the ordered α'-phase and the other is the ordered γ-phase (the space group C2/m). Thus, the partially ordered α'-phase in the $TiN_{0.26}H_{0.15}$ solid solution is metastable. In the range of 1170 ÷ 870 K a chain of phase transitions is observed:

$$\text{Disorder} \xleftrightarrow{1170K} \text{Short-range order} \xleftrightarrow{\approx 1020K} \text{Long-range order} \xleftrightarrow{870K} \text{Disintegration}$$

or in terms of phases:

$$\alpha \xleftrightarrow{1170K} \alpha \text{ with short - range order} \xleftrightarrow{1020K}$$

$$\alpha' \xleftrightarrow{870K} \alpha' + \gamma.$$

The above pattern makes it clear that the decay precedes by (partial) long-range order, and that long-range order precedes by short-range order. However, this is only a simplified model. Actually, the temperature intervals of the existence of the short-range order, the long-range order and the decay overlap. The short-range order exists in the range of 1170 ÷ 820 K; the long–range order exists between 1020 and 740 K; the shot-range order coexists with the long - range order at 1020 ÷ 820 K; and the order is changed by the decay at 850 ÷ 800 K. The dominant process is a function of the temperature. The rate at which ordering and decay processes alter each other depends on temperature too. If at 850 K features of the decay are observed after exposure of 18 h, at 800 K they appear only after 48 h. To shed light upon the ordering process in the $TiN_{0.26}D_{0.15}$, the disordered samples were annealed at various temperatures in the range of temperatures 1190 ÷ 740 K, and the state corresponding to each temperature was frozen by quenching in air. Based on the neutron data (Fig. 2a), the sample of the $TiN_{0.26}D_{0.15}$ solid solution quenched from 1270 K have a hexagonal structure (the space group $P6_3/mmc$, the L'_3 type) which the disordered $TiN_{0.26}H_{0.15}$ had [9]. Processing of the neutron diffraction pattern of the disordered solid solution showed that the nitrogen atoms statistically occupy the octahedral interstices 2(a), the deuterium atoms –both the octahedral 2(a) and the tetrahedral ones 4(f) in the space group $P6_3/mmc$ (α phase). The neutron diffraction pattern of the $TiN_{0.26}D_{0.15}$ quenched from 1190 K (Fig. 2b) or 1160 K for 5 h display superstructure reflections which are the same as in the neutron diffraction pattern of the ordered $TiN_{0.26}H_{0.15}$, its structure is described in the space group $P\bar{3}m1$ (α′ phase). Therefore, the crystal structures of the ordered $TiN_{0.26}D_{0.15}$ and $TiN_{0.26}H_{0.15}$ are of the same space group $P\bar{3}m1$. The X-ray powder diffraction shows that the quenched solid solutions are single-phases with the same unit cell parameters as that of the disordered solid solution. Two circumstances are noteworthy: firstly, notable superstructure reflections in the $TiN_{0.26}D_{0.15}$ neutron diffraction pattern appear at a higher temperature than in the $TiN_{0.26}H_{0.15}$ (~1190 K against ~ 1020 K [5]); secondly, the neutron diffraction pattern of the $TiN_{0.26}H_{0.15}$ sample, quenched from these temperatures, shows a diffuse reflection (indicative of short-range order [9]) which was not observed for the deuterium solid solutions of the same composition. Next, quenches from 1070 K and 1020 K strengthen the superstructure reflections in the $TiN_{0.26}D_{0.15}$ neutron diffraction pattern, but a diffuse reflection does not appear. It should be noted that the study of short-range order in the deuterium solid solutions is difficult in comparison with the hydrogen one. The matter is that the diffuse reflection intensity of solid solutions in the Ti-N-H and Ti-N-D systems is proportional to the squared difference between the neutron scattering amplitudes for nitrogen and hydrogen: $I \sim (b_N - b_H)^2$ [20].

It is clear that for $b_N = 0.94 \times 10^{-4}$, $b_H = 0.374 \times 10^{-4}$, $b_D = 0.667 \times 10^{-4}$ Å [13], the substitution of the neutron scattering amplitude of deuterium instead of that of hydrogen essentially decreases the $(b_N - b_{H(D)})^2$ value and, accordingly, reduces the diffuse reflection intensity for solid solutions in the Ti-N-D system. The Rietveld analysis of the neutron diffraction pattern of the $TiN_{0.26}D_{0.15}$ quenched from 1190 K shows that a good agreement between the measured and calculated peak intensities (Fig. 2b) and the lowest $R_{Br} = 5\%$ is achieved for the following atom arrangement in the space group $P\bar{3}m1$: two titanium atoms occupy the 2(d) positions with the coordinates 1/3, 2/3, z_{Ti} and 2/3, 1/3, $\bar{z}_{Ti}$ ($z_{Ti} = 0.248 \pm 0.002$); 67% of the nitrogen atoms occupy the octahedral interstices 1(a) and 33% of them occupy the octahedral interstices 1(b); 50% of the deuterium atoms are located in the tetrahedral interstices 2(d) with a variable the parameter $z_D = 0.603 \pm 0.006$ and 50% are in the octahedral interstices 1(b). Therefore, the structure model is an alternation of layers built on metal atoms and two types of octahedrons and tetrahedrons. Note that the tetrahedral interstices 2(d) with z_D = 0.128 lying near the nitrogen plane remain vacant. If to suppose that deuterium atoms are arranged only over two types of tetrahedrons, the condition R_{Br} < 14% was unachievable. The experimentally derived long-range order parameter [21] for the N sublattice η_{expN} = 0.17 is one-third of the maximum value η_{maxN}=0.52. Thermal processing was continued to achieve the maximum ordering. The processing of the neutron diffraction pattern of the same sample after annealing it at 1020 K for 5 h increased the long-range order parameter η = 0.23. The intensity of the superstructures on the neutron diffraction pattern increase, especially that of the superstructure 001. No changes are observed in the X-ray diffraction patterns.

To obtain equilibrium state, the $TiN_{0.26}D_{0.15}$ was annealed for 24 h at temperature 1020 K. The nitrogen the long-range order parameter increased to η^{N}_{exp}=0.36; 57% of the deuterium atoms occupy the tetrahedral interstices 2(d) and 43% reside in the octahedral 1(b) (Table 1). Arrangement of the deuterium atoms in the octahedral interstices 1(b) transfers most the nitrogen atoms to transfer to other the octahedral ones 1(a). This is unambiguous evidence that some deuterium atoms remain in the octahedrons unoccupied by nitrogen atoms.

This character of deuterium distribution differs from hydrogen distribution in the $TiN_{0.26}H_{0.15}$ solid solution: in the latter, the hydrogen atoms at the initial ordering stage occupy two types of tetrahedrons [9, 12]. It is worth noting the value of the deuterium free parameter z_D in the ordered $TiN_{0.26}D_{0.15}$. In the disordered $TiN_{0.26}D_{0.15}$, the value z_D (equal to 0.635) was larger than that of the tetrahedron center ($z_{tetr.}$=0.628) [12]; in the ordered solid solution, z_D< $z_{tetr.}$(Table 1).

Consequently, an alteration in the character of interstitial nitrogen distribution changes the sign of the displacement of deuterium atoms from the tetrahedron center. However, the coordinate of metal atoms along the *c*-axis z_{Ti},, coincides (within the experimental errors) with the value of the free parameter for an ideal hexagonal close packing lattice (z_{id}= 1/4), in contrary to the ordered $TiN_{0.26}H_{0.15}$ solid solution (for which z_{Ti}= 0.243).

Then we studied the effect of subsequent low-tempera-ture anneals on the structure of the ordered $Ti_2N_{0.52}D_{0.30}$ (α'-phase). First, partially the ordered $Ti_2N_{0.52}D_{0.30}$, prepared by annealing at 1190 K and quenched, was annealed at lower temperature 900 K for 24 h.

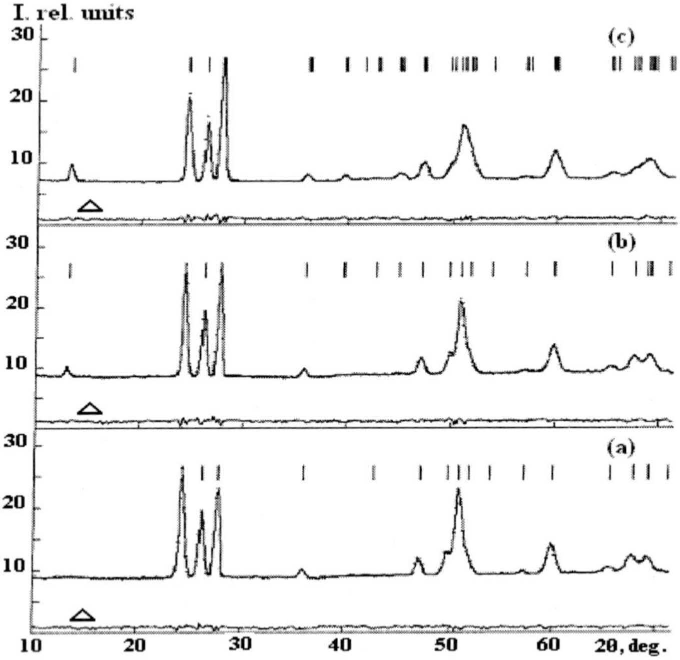

Figure 2. Neutron diffraction patterns of the $TiN_{0.26}D_{0.15}$ annealed at 1270 K (a), 1190 K (b) and 900 K (c): notations are the same as in Fig. 1.

Table 1. The structure parameters of the $TiN_{0.26}D_{0.15}$ annealed at the temperature 1020 K for 25h (space group P3ml)

Atom	Position	x/a	y/a	z/c	n
Ti	2d	1/3	2/3	0.25_2	2.000
N1	1a	0	0	0	0.34_7
N2	1b	0	0	½	0.17_3
D1	2d	1/3	2/3	0.60_3	0.16_1
D2	1b	0	0	½	0.13_9
B_{ef}, Å^2	B_{ef}=0.53±0.05				
R, %	R_P = 1.6; R_{wp} = 2.3; R_{Br} = 4.5				

Note B ef. Is the effective thermal factor, n is the site occupance factor, R is the discrepancy index.

This thermal processing conserved the extant neutron diffracttion peaks but increased the half-width of structure reflections; the half-width of superstructural reflections remained unchanged (Fig. 2c). At the same time, because of high resolution, a number of the X-ray diffraction peaks undergoes clear splitting (Figure 3).

The character of splitting was identical to that of observed for the monoclinic γ-phase with the structure type of anti - $AuTe_2$ which has been observed in the $TiN_{0.32}H_{0.19}$ and $TiN_{0.40}H_{0.21}$ solid solutions [8]. Both the X-ray and the neutron diffraction patterns are indexed in monoclinic unit cell with lattice parameters $a \approx \sqrt{3}\, a_0$, $b \approx a_0$, $c \approx c_0$, $\beta \approx 89.73^0$ where a_0 and c_0 are parameters of the initial hexagonal cell. Good agreement between the observed and calculated peak intensities (Fig. 2c) and the least discrepancy indexes (Table 2) were achieved in the space group C2/m for the following arrangement of intersttial atoms: the nitrogen atoms occupy one type of the octahedral interstices 2(a) of two; 73% of the deuterium atoms occupy the tetrahedral interstices 4(i) and 27% −the octahedral interstices 2(c). The coordinate z of metal atoms remains ideal: $z_{id} = 1/4$ (within the calculation errors). Locating nitrogen and deuterium atoms in other positions noticeably deteriorates the agreement with the model. Note, that in the monoclinic phase the character of atomic ordering remains the same as in the hexagonal phase of this solid solution. Consequently, the observed transition is effect of thermal elastic stresses on the ordered phase structure and is due to the host lattice distortion. Annealing at 740 K for 48 h does not lead to ordering of metalloids in the $TiN_{0.26}H_{0.15}$ and $TiN_{0.26}D_{0.15}$ solid solutions.

Ordering occurs only after annealing during 186 h. The X - ray diffraction pattern of these samples annealed at 740 K does not show features of decay even after 186 h of exposure. The neutron diffracttion pattern with smaller scan step of 5' revealed splitting of superstructure reflections (Figure 4). Splitting of each of these reflections into a triplet of symmetrically arranged reflections is evidence that a periodic, ordered antiphase domain structure is formed below the decay point (at $T<T_d$) [22]. It should be noted that the splitting is observed not only for 00*l* superstructure reflections but also for 111 reflection (which is superstructure one also). Therefore, the domain ordering in the solid solution has no layered character. The intersatellite distance was used to determine the period of the ordered antiphase domain structure. The period determined from the 003 reflection is ~10*c*.

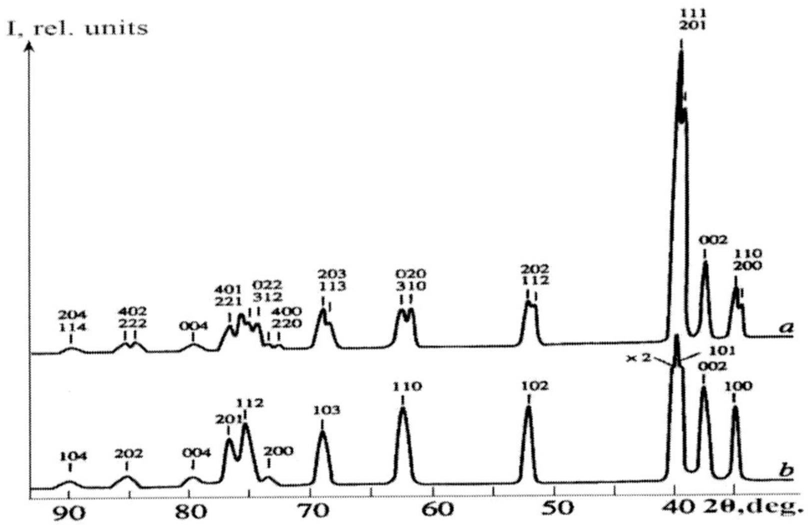

Figure 3. X-ray diffraction patterns of the $TiN_{0.26}D_{0.15}$ sample annealed at 1270 K (*a*) and 900 K (*b*). Above the peaks Miller indexes of the reflecting planes are indicated.

Table 2. The structure parameters of the $y\text{-}Ti_2N_{0.52}\,D^{0.30}$ phase (space group C2/m)

Atom	Position	*x/a*	*y/b*	*z/c*	n
Ti	4 *i*	0.317_9	0	0.249_5	4.000
N	2 *a*	0	0	0	1.040
D1	4 *i*	0.349_6	0	0.609_1	0.44_6
D2	2 *c*	0	0	½	0.16_6
$B_{ef.}$, Å^2	0.26 ±0.03				
R, %	$R_P = 1.6$; $R_{wp} = 2.3$; $R_{Br} = 4.5$; $R_F = 4.7$				

The absence of antiphase domain ordering in the samples annealed at high temperatures can be considered as a formation of large and randomly oriented antiphase domains at these temperatures. Fig. 5 shows the neutron pattern of the $TiN_{0.26}H_{0.075}D_{0.075}$ obtained by quenching from 1270 K.

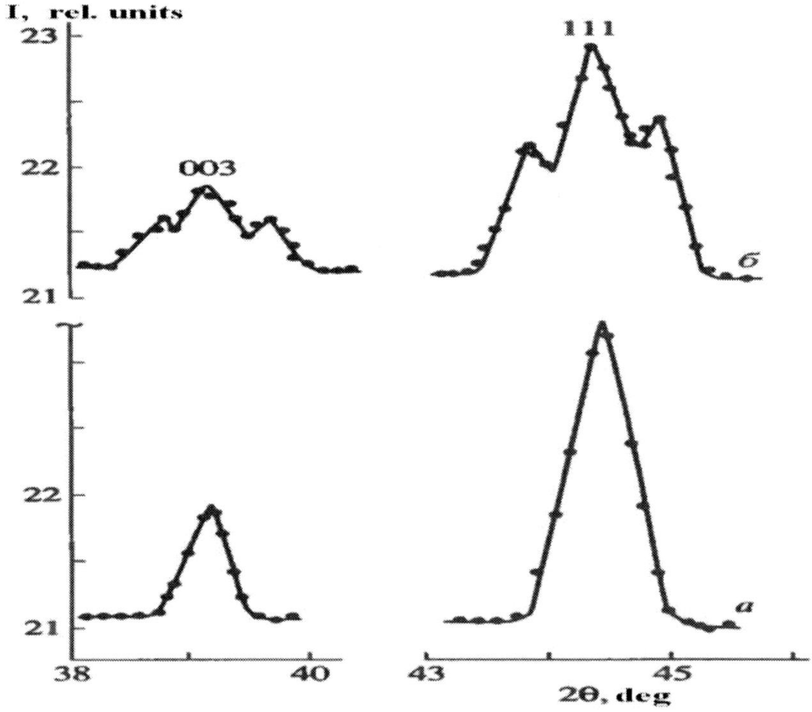

Figure 4. The superstructure reflections on the neutron diffraction pattern of the $TiN_{0.26}H_{0.15}$ after annealing for 186 h at 890 K (a) and 740 K(b).

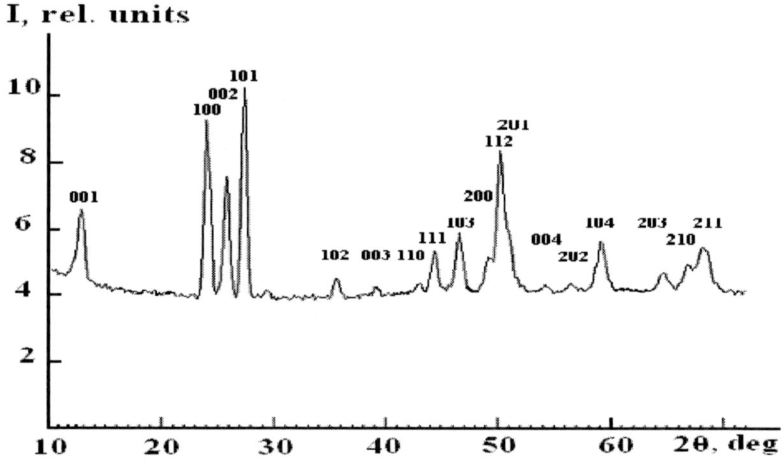

Figure 5. Neutron diffraction pattern of the sample $TiN_{0.26}H_{0.075}D_{0.075}$ quenched from 1270 K.

According to the neutron data, the $TiN_{0.26}H_{0.075}D_{0.075}$ solid solution has a hexagonal ordered structure (the space group $P\bar{3}m1$) were nitrogen atoms occupy the octahedral interstices 1(a), hydrogen atoms are located in the tetrahedral interstices 2(d) with $z = 0.754$; 80 % of the deuterium atoms are in the tetrahedral interstices 2(d) with $z = 0.636$ and 20 % – in the octahedral interstices 1(b). Unlike the $TiN_{0.26}H_{0.15}$ and $TiN_{0.26}D_{0.15}$ solid solutions (Figure 1 and Figure 2a, accordingly), $TiN_{0.26}D_{0.075}H_{0.075}$ in the high-temperature state has an ordered structure (Fig.5). Hence, the temperature of order-disorder transition in the $TiN_{0.26}H_{0.075}D_{0.075}$ is higher than in the $TiN_{0.26}H_{0.15}$ and $TiN_{0.26}D_{0.15}$ solid solutions. The neutron diffraction study of the $TiN_{0.26}H_{0.15}$, $TiN_{0.26}D_{0.15}$, $TiN_{0.26}H_{0.075}D_{0.075}$ solid solutions and comparison of the obtained results with results of of investigation of the $TiN_{0.26}H_{0.15}$ allowed one to reveal the following isotopic effect:

1. Nitrogen ordering in the $TiN_{0.26}D_{0.15}$ and $TiN_{0.26}H_{0.075}D_{0.075}$ (space group is $P\bar{3}m1$) starts at higher temperatures (~1270 K and ~1170 K, accordingly) than in the $TiN_{0.26}H_{0.15}$ (~1020 K).
2. Unlike the $TiN_{0.26}H_{0.15}$, in the $TiN_{0.26}D_{0.15}$ diffuse scattering does not precede the appearance of superstructural reflections in the neutron diffraction patterns.
3. The low-nitrogen homogeneity limit of the monoclinic ordered phase (the space group is C2/m) in the Ti-N-D system lies at lower nitrogen concentrations than in the Ti-N-H system.
4. Hydrogen atoms in the ordered $TiN_{0.26}H_{0.15}$ arrange only over tetrahedral interstitial; in the ordered $TiN_{0.26}H_{0.075}D_{0.075}$ and $TiN_{0.26}D_{0.15}$ solid solutions some deuterium atoms can also occupy the octahedral interstices unoccupied by nitrogen atoms.
5. In the ordered α'- and γ- phases of the T-N-D system, as opposed to the same phases of the T-N-H system, the z coordinate of the metal atom remains ideal to within the calculation error ($z_{\mathrm{Ti}} \approx 0.25$), whereas in the monoclinic γ - phase, the x_N and x_{Ti} coordinates deviate significantly from the ideal value of $x_{id} = 1/3$.

Evidently, these differences are due to the isotopic features of hydrogen and deuterium atoms.

The higher temperature of ordering in the Ti-N-D than in the Ti-N-H system can be explained as follows. In imperfect solid solutions, strains and long-range interact-tions play an essential role. One can therefore suggest that substitution of H by D atoms in the solid solutions increases the lattice strain field and, probably, enhances the stability of the ordered state. An example of the $TiN_{0.26}H_{0.075}D_{0.075}$ shows that nonhomogeneity in the sublattice of interstitial atoms else raises the deformation interaction and, accordingly, results in increasing the temperature of order-disorder transition. As regards the location of part of the D atoms in octahedrons of ordered α'- and γ-phases, it may be supposed that the blocking effect caused by repulsion forces between a nitrogen atom in 1(a) octahedron and deuterium atom in tetrahedron is stronger than between N and H atoms. As a result, part of deuterium atoms arrange at longer distances (in 1(b) octahedron): $d_{N\text{-}D2}$=2.569>$d_{N\text{-}D1}$ = 2.413 Å.

The crystal structure of the solid solution with the combined isotope composition $TiN_{0.40}H_{0.19}D_{0.19}$ has been studied by the neutron and X-ray diffraction methods [19].

According to the X-ray diffraction patterns, the synthesized $TiN_{0.40}H_{0.19}D_{0.19}$ solid solutions was single-phase and had the hexagonal (trigonal) unit cell with the parameters a = 3.009, c = 4.875 Å The neutron diffraction pattern of the $TiN_{0.40}H_{0.19}D_{0.19}$ solid solution presented in Fig. 6. The neutron diffraction pattern of the single-phase and homo-geneous (on composition) $TiN_{0.40}H_{0.19}D_{0.19}$ solid solution, as of $TiN_{0.40}H_{0.21}$, was indexed in the space group $P\bar{3}m1$.

The Neutron diffraction data can be explained well, only if to assumed that all nitrogen atoms occupy the octahedral interstices 1(a) and atoms H and D - the tetrahedral interstices 2(d) but the last with different value of coordinate z (Tab. 3).

If assumed that the isotopes H and D would have identical coordinate z, then the agreement between experimental and calculated intensities sharply decreases, particularly at the small Bragg angles (where the influence of thermal factor is still insignificant) and, respectively,the divergence factor R_{Br} rises from 6 % up to 11%. Thus, in contrary to the ordered TiN_xH_y [6, 9] and TiN_xD_y solid solutions [12], in solid solution with combined isotopic composition $TiN_xH_{y/2}D_{y/2}$, the tetrahedral interstices 2(d) are split into the proton and deuterium positions.

Isotope ordering of hydrogen isotopes over octahedral and tetrahedral caused by zero vibrations of the isotopes was observed earlier in the hydride CeD_2H at low temperatures T < 300 K [14]. The isotope distribution (ordering) of the hydrogen isotopes over interstices of the same type in a single-phase high-temperature state of compounds is found for the first time.

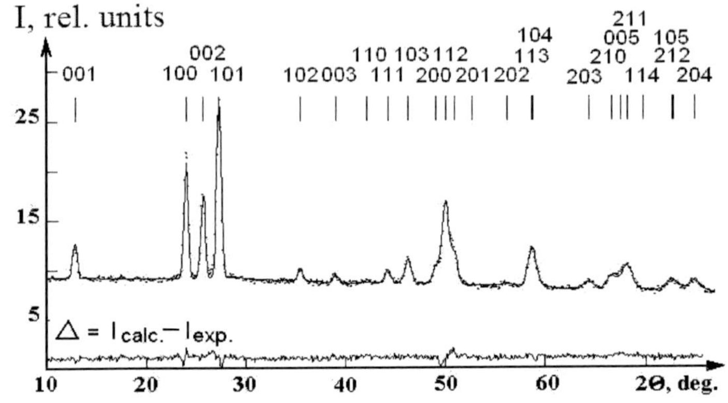

Figure 6. Neutron diffraction pattern of the solid solution $TiN_{0.40}H_{0.19}D_{0.19}$: solid line is calculated, dots are experimental, Δ is differential ones.

Table 3. The structure characteristics pf the TiN $_{4.40}$ H $_{0.29}$ D $_{0.19}$ solid solution in the frame of space group P3ml

Atom	Posi-tion	x/a	y/a	z/c	Δz	n	Δn
Ti	2 d	0.333	0.666	0.248	0.002	2.000	0.000
N	1 a	0.000	0.000	0.000	0.000	0.800	0.000
D	2 d	0.333	0.666	0.582	0.004	0.316	0.015
H	2 d	0.333	0.666	0.794	0.007	0.327	0.028
B_{ef}= 0.73±0.04 Å; R_P=2.07; R_{WR}=2.69; R_{exp}=1.51; R_{Br}=6.69 %							

For more detailed description of the mentioned solid solutions it is convenient to use crystal-chemical approach: representation of a structure as a combination of Pauling polyhedrons on the basis of the closest packing of solid spheres. According to [23], atomic radii of Ti, N and H are equal to 1.46; 0.71 and 0.46 ÷ 0.37 Å, respectively. The atomic radius of deuterium may be considered as close to 0.50 Å. Based on the sizes of atomic radii and concentration of elements, the large atoms of titanium are assumed as a basis of structures, i.e. two-layer spherical packing in octahedral and tetrahedral interstices of which the rest atoms are arranged. According to [23], the radius of octahedral interstices is ~ 41 % and of tetrahedral one - ~ 22 % of a radius of the basis solid sphere. It gives $r_{oct} \approx 0.60$ and $r_{tetr} \approx 0.32$ Å, respectively, i.e. the size of octahedrons gives the fit for arrangement of N, and tetrahedrons - for H (D). It is convenient to describe the structure as a set of octahedrons and tetrahedrons similarly to the way used in [8]. Based on the atoms coordinates determined by the neutron diffraction data for the studied solid solution (Table 2), it was constructed the regular system of points using formulas of multiplication for the space group $P\bar{3}m1$ [23]. For each polyhedron the distances from the central atom (N, H and D) up to the tops (atoms Ti) and also the polyhedron edges (distances Ti – Ti) were calculated. The distance $r_{1\text{-}2}$ between atoms 1 and 2 is defined by the formula for hexagonal system:

$$r_{1\text{-}2} = (x_2 - x_1)^2 + (y_2 - y_1)^2 + (z_2 - z_1)^2 - (x_2 - x_{x1})(y_2 - y_1),$$

where x_n, y_n and z_n are coordinates of corresponding atoms. Calculations became simpler taking intoaccount that the atoms Ti, H and D are in the particular positions with multiplicity 2 - on the threefold axes and so have coordinates: (a/3, 2a/3, zc) and (2a/3, a/3, -zc), and the N atom is in the center of symmetry, i.e. it has multiplicity 1. The results are shown in Tab. 4. For comparison the corresponding distances in the structure of the ordered $TiN_{0.40}H_{0.21}$ solid solution [8] are also given.

N–octahedron. Table 4 shows that the sizes of N-octahed-rons corresponds to the crystal-chemical criteria: $r_{N\text{-}Ti} \approx 2{,}1$ Å, $r_{Ti\text{-}Ti} \approx 3$Å.

Table 4. Interatomic distances r(A) in the solid solutions

Bond	r, Å	
	$TiN_{0.40}H_{0.19}D_{0.19}$ (According to the data of [19])	α'-$TiN_{0.40}H_{0.21}$ (According to the data of [8])
	N – octahedron	
N – Ti	2,120	2,098 (N_1) 2,136 (N_2)
$Ti_1 - Ti_1'$	3,014	3,012
$Ti_1 - Ti_2$	2,983	2,921 (N_1) 3,030 (N_2)
	D – tetrahedron	
Ti_1-D	1,629	
Ti_2-D	1,926	
Ti_2-Ti_2'	3,014	
Ti_1-Ti_2	3,008	
	H-tetrahedron	
Ti_1' - H	2,214	
Ti_1 – H	2,663	1,850
Ti_2 – H	1,753	1,850
Ti_2- Ti_2'	3,014	3,012
Ti_1– Ti_2	2,983	3,030

Since a threefold inversion axis passes through the center of octahedron, the pair of faces, which are the regular triangles ($r_{Ti\text{-}Ti} = a$), is placed in parallel with the basic plane (001) (Figure 7) and, thus, joining by the common faces, octahedrons form a layer of height $2z_{Ti}c$ parallel to the (001) plane. This is the layer, in octahedral interstices, in which the nitrogen atoms are arranged. The second layer of octahedrons, adjacent on height, is empty, its height is equal to $(1\text{-}2z_{Ti})c$; the difference is insignificant: height of the filled layer is 0,4968c (~2.42 Å) and of the empty one – 0.5032c (~2.46 Å). In the interspaces between empty octahedrons of the unfilled layer the tetrahedrons are placed in which the deuterium and the hydrogen atoms are located. By virtue of the well- known effect of H-H blocking [24], the D and the H atoms cannot be arranged simultaneously in the same tetrahedron as it formally follows from Table. 3. Therefore, the D and the H tetrahedrons are considered as independent units.

D – tetrahedrons. In accordance with radii, the distance r(D –Ti) ≈ 2 Å. So it should be noted the fact (Tab. 4) that one of the distances up to corner (Ti) is shortened to 1.629 Å (for comparison: the distance Ti – H(D) in titanium dihydride is 1.93 Å, in saturated hydrides of the Ti inter-metallic compounds – 1.87÷1.99 Å; in solid solution of hydrogen in titanium – 1.803 Å). Thereby the deuterium atom is displaced from Ti -"bottom" of a tetrahedron towards a top (Figure 8), i.e. towards the layer of N-octahed-rons. Such a shift is likely caused by the interacttion forces between the interstitial D and H atoms, at least, in the first coordination sphere.

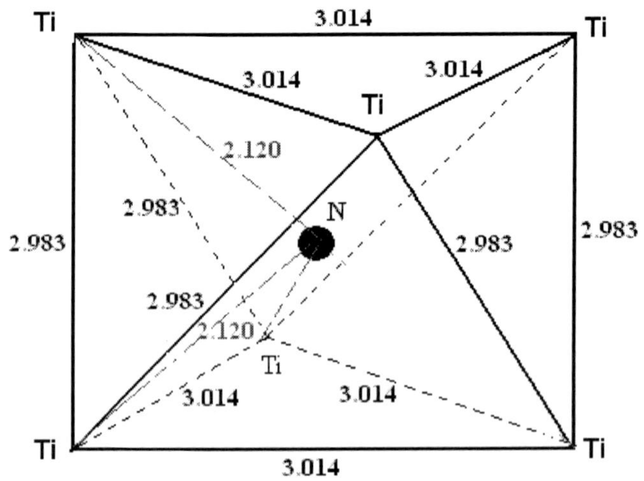

Figure 7. N – octahedron. Basis interatomic distances are given in Å.

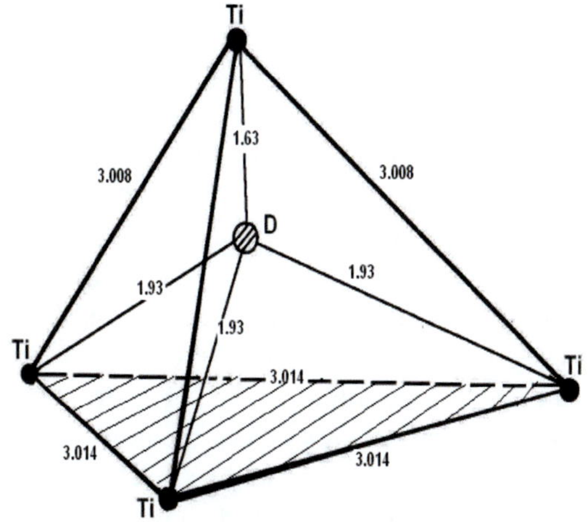

Figure 8. D – tetrahedron; the base parallel to (001) is hatched. Interatomic distances are given in Å.

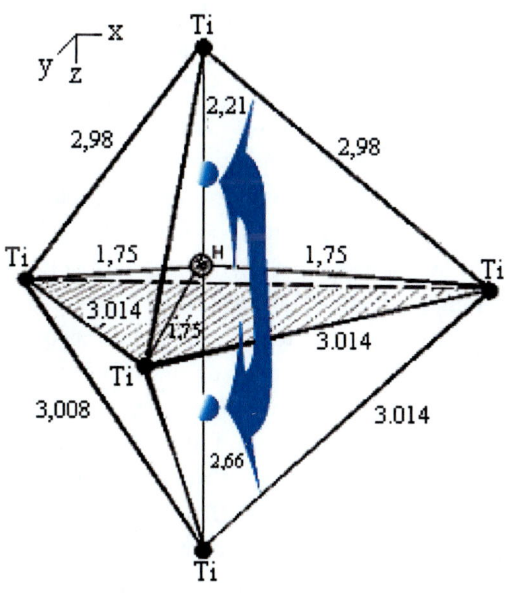

Figure 9. Coordination of H atom; the common base of two adjacent pyramid is hatched. Arrow and light balls symbolically indicate action of resonance mechanism. Interatomic distances are been given in Å.

H – polyhedron. With H – polyhedron (tetrahedron?) the situation is more intricate. First of all, if D-tetrahed-rons are located in a layer of empty N -

octahedrons, H -"tetrahedrons" are in the occupied layer (formally). At the theoretical value r(H-Ti) ≈ 1.9 Å, very big difference in the sizes of distances H –Ti for various Ti atoms (Table 4, Figure 9). One can see that the distances from the atoms H to the atoms Ti_2 which form the base of a smeller tetrahedron (i.e. parallel to the (001) plane), are identical and are much than distances for apical atoms both of "own" and adjacent tetrahedrons (along the axis c) having the same base. The hydrogen atom is at a very small distance from this common base – 0.0429c (~0.21Å). The values of the H–Ti distances in the $TiN_{0.40}H_{0.19}D_{0.19}$ it can be conclude that the coordination of the hydrogen atom is changed: in the first coordination sphere there are 3 Ti atoms at the distances about 1,7 Å; at the second sphere there are 2 Ti atoms (in apical position) at the distance 2.2÷2.7 Å. It should be noted that the distance from H to Ti- top in the "own" tetrahedron is larger than in adjacent one (2.663 against 2.214 Å). It leads to conclusion that the coordination of hydrogen in this solid solution should be regarded as threefold (triangular). In crystal chemistry such coordination is frequently observed for atoms with small sizes; for example, it is usual for the boron atoms [25]. In hydride systems, this effect is found for the HoD_3, YD_3 and Th_4D_{15} compounds [26]. If the atoms from the second coordination sphere included into the polyhedron, then one can see the united base and two tops lying at the opposite sides from it, so, it will be found unusual for such structures polyhedron - triangular bipyramid in the middle of which the H-atom is placed (Fig. 9). Such coordination has been occurred only in solid solutions with combined isotope composition of hydrogen. However taking into account energy aspect, it seems less probable that the hydrogen atoms prefer to arrange not in tetrahedrons (and octahedrons) but in small triangular interstices (if, certainly, to exclude occurrence of "chemism", i.e. chemical bonds of TiH_2 type is excluded) at the large quantity of empty polyhedrons (amount of tetrahedrons is twice as large as amount of octahedrons) and at relatively low concentration of the hydrogen atoms. It is more probably that the peculiarities of structure analysis methods have an important influence upon the results. Diffraction spectrum is some statistic information averaged over all irradiated volume. Therefore, it may assume that actually the small atoms of hydrogen can both statistically arrange over usual tetrahedrons and also migrate from one interstice another, due to the well-known resonance (swing) mechanism of oscillation [25]. Note that a pair of tetrahedrons having common base in hexagonal close packing lattice of the $TiN_{0.40}H_{0.19}D_{0.19}$ solid solution can not be occupied by H atoms simultaneously since their similar arrangement leads to inadmissibly short distance H – H (from viewpoint of blocking effect). Neutron diffraction experiment gives coordinates of some average position of H-atoms, i.e. an approximate center of triangular bipyramid which we observe in practice. Asymmetry of the mentioned middle position relative to the common base of tetrahedrons may be explained by non-equivalence of real hydrogen positions “upper” and “lower” tetrahedrons and also by various residence time of hydrogen atom in these tetrahedrons. Therefore, in the $TiN_{0.40}H_{0.19}D_{0.19}$ solid solution the third interstitial element (N) occupies octahedron of one of two parallel layers but octahedron of the second layer is empty. The atoms H and D (of smaller size), concentration of which is insignificant, are placed not in empty octahedrons but in tetrahedrons. Therefore it is possible to assume that deuterium occupies tetrahedrons located in the layer of empty octahedrons being displaced from the geometrical centre of a tetrahedron toward the N-layer. The hydrogen most likely, migrates both over adjacent tetrapores of filled and empty layer of octahedrons. Therefore, unlike the CeD_2H alloy [14], where isotopes arrange over interstices of different types: in octahedral and tetrahedral interstices with excess of light isotope in octahedron interstices. In the given solid solution the isotopic splitting of the same type interstices 2(d) takes place. Apparently, it is possible to explain this difference by the fact that in the $TiN_{0.40}H_{0.19}D_{0.19}$ solid solution interaction forces between interstitial atoms (N, H and D) is noticeable, unlike CeD_2H. A deformation interaction between interstitial atoms prevails rather than electronic that leads to the arrangement of isotopes H and D at different distances from other atoms. In this case, in the ordered "layered" hexagonal structure of the $TiN_{0.40}H_{0.19}D_{0.19}$ solid solution, two isotopes (H and D) of hydrogen cannot be located at the same plane parallel to the (001) plane (perpendicular to the ***c***-axis).

CONCLUSION

The high-temperature (1270 K) disordered $TiN_{0.26}H_{0.15}$ and $TiN_{0.26}D_{0.15}$ solid solutions of similar composition have the identical structure characteristics: their unit cell parameters are similar, and the arrangement of nonmetal atoms in interstices are identical. However, an isotopic effect appears when the temperature is decreased. This means,that in the $TiN_{0.26}D_{0.15}$, unlike the $TiN_{0.26}H_{0.15}$, the process of ordering begins at higher temperature; detection of a short-range order is difficult; the monoclinic γ-phase has much greater homogeneity range; in the ordered α- and γ-phases small part of deuterium atoms are in octahedrons. The blocking effect between interstitial atoms in the deuterium solid solution is more pronounced than in the hydrogen solution, which is evidence in favor of the prevalence of strain interactions between hydrogen atoms and other atoms. The ordering of anti-phase domains is been found out in these solid solutions.

It is shown that in a number of the $TiN_{0.26}H_{0.15} \rightarrow TiN_{0.26}D_{0.15} \rightarrow TiN_{0.26} H_{0.075}D_{0.075}$ solid solutions the tem-perature of ordering raises as 1025 K→1175 K→ 1275 K, accordingly. It is possible to explain this process by the fact that in the strongly defective solid solutions the deformation interaction on the interstitial atoms plays an essential role. Apparently, the replacement of H and by D in a solid solution leads to strengthening of deformation interaction and to the temperature increase of the order-disorder transition. The nonhomogeneity of the hydrogen sublattice (in the $TiN_{0.26} H_{0.075}D_{0.075}$) increases deformation interaction. This raises the temperature of ordering.

The location of hydrogen atoms in the disordered $TiN_{0.26}D_{0.15}$ solid solution showed that they are displaced from the center of the tetrahedral array of metal atoms away from the nearest nitrogen atoms. Therefore, repul-sion force between exists between the nitrogen and the hydrogen atoms. In an ordered solid solution, the dis-placement of a hydrogen atom from the tetrahedral center changes its sign in correlation with the redistribution of nitrogen atoms and vacancies.

The crystal structure of the solid solution with combined isotopic composition $TiN_xH_{y/2}D_{y/2}$ is investigated for the first time. The isotopic effect in the distribution of the hydrogen isotopes H and D is found out; this means that the isotopes H and D arrange in one type of the tetrahedral interstices 2 (d) with different coordinates z, that is the splitting of the tetrahedral interstices 2d to hydrogen (H) and deuterium (D) ones is observed. It follows that in the ordered "layered" hexagonal structure (the space group $P\bar{3}m1$) of the $TiN_xH_{y/2}D_{y/2}$ solid solutions there can not be two different hydrogen isotopes on one plane (parallel to the (001) plane).

Apparently, in the given solid solutions deformation interaction between hydrogen atoms and other atoms prevails more likely than electronic one that results in an arrangement of isotopes H and D at the different distances from other atoms.

The author is grateful to Dr. M. J. Ermamatov and Dr. E. Tursunov for the revision of the Chapter and useful comments.

REFERENCES

[1] Kleefisch, E. W. *Industrial Applications of Titanium and Zirconium*; PA: ASTM International: West Conshohochem, USA, 1981; 286 p.

[2] Kornilov, I. I. *Titanium; Nauka: Moscov, Russia*, 1975; 308 p.

[3] Shveykin, G. P., Alyamovskiy, S. N., Zaynulin, Yu. G. et al. *Compounds of variable composition and their solid solutions*; Ural Center of Sciences of AS USSR: Sverdlovsk, Russia, 1984; 294 p.

[4] Lobier, G. Contribution a l'etude de l'insertion de l'hydrogen dans les carbures, nitrures et sulfures des elements de transition.*Rapport Commis. Energ. At. Paris.* 1970. № R-4065. 133 p.

[5] Sumin, V. V., Morozov, S. I. *Russian J. Solid State Physics (Rus).* 1987, v.63, 1159-1164.

[6] Khidirov, I., Karimov, I., Em, V. T. et al. *Russian J. Inorg. Mat.* (Rus).1981, v. 17, 1416-1420.

[7] Morozov, S. I. Investigation of admixture fluctuations of atoms of introduction in zirconium and titanium by method of inelastic dispersion of slow neutrons: Cand. dissertation (phys.-math. sci.). *Obninsk*, 1989.194 p.

[8] Khidirov, I., Kurbonov, I. I, Mukhtarova, N. N. *Kristallografia (Rus).* 1996, v. 41, 450-456.

[9] Khidirov, I. Processes of ordering and decay in the nitrogen and hydrogen solid solution in α - Titanium/ *Russian J. Inor. Chem. (Rus).* 2001, v. 46, 499-505.

[10] Matysina, Z. A., Schur, D. V. *Hydrogen and solid-phase transformations in metals, alloys and fullerits; Nauka i obrazov.*: Dnepropetrovsk, Ukraine, 2002; 420 p.

[11] Khidirov, I., Kurbonov, I. I., Padurets, L. N. *Metallofizika (Rus).* 1993, v. 15, 87-90.

[12] Khidirov, I., Paduretc, L. N. *Russian J. Inor. Chem. (Rus.).* 2001, v. 46, 1413-1418.

[13] Bacon, G. E. *Neutron Diffraction*; Clarendon Press: Oxford, England, 1975; 636 p.

[14] Somenkov, V. A., Entin, I. P., Kost, M. E. et al. *Russian J. Solid State Phys. (Rus).* 1975, v. 17, 2368-2372.

[15] Shoyusupov, Sh., Getmanskiy, V. V., Khidirov, I., et al. *Uzbek J. Problem of energetics and automatics.* 2002, no 2, 11-16 (in Rus.).

[16] Young, R. A., Wilas, D. B. Profile Scope Functions in Rietveld Refinements. // *J. Appl. Cryst.,* 1982, v. 15, 430-438.

[17] Amosov, A. P., Borovinskaya, I. P., Merzhanov, A. G. Powder technology of self-propagating high-tempera-ture synthesis of materials; *Mashinostroenie -1: Moscow, Russia*, 2007; 471 p.

[18] Khidirov, I., Padurets, L. N., Zaginaichenko S. Yu., et al. Proc. of the NATO Advanced Research Workshop on Hydrogen Materials Science and Chemistry of Carbon Materials, Sevastopol, Ukraine, 5-11 September 2005. *Hydrogen Materials Science and Chemistry of Carbon Nanomaterials;* Edited by Veziroglu, T. N., Zaginaichenko, S. Yu.,et al.; Springer: Dordrecht, The Netherlands, 2007. PP. 65-75.

[19] Khidirov, I., Mukhtarova, N. N., Padurets, L. N. , et al. *Russian J. Inorg. Chem. (Rus.).* 2008, v. 53, 564-569 (in Russian).

[20] Khvatinskaya, D. Ya, Kurbonov, I. I, Khidirov, I, et al. *Uzbek J. of Phys.1991*; no 3, pp. 67-70 (in Russian).

[21] Gusev, A.I. Nonstoichiometry, disorder, short-range order and long-range order in solid state. *FizMatLit: Moscow, Russia*, 2007; 416 p.

[22] Toth, R. S., Sato, H. *J. Appl. Phys.* 1962; 33(11): 3550-3256.

[23] Bokiy, G. B. *Kristallokhimiya. Nauka: Moscow, Russia*, 1971; 400 p.

[24] Shilov, A. L., Padurets, L. N. *Russ. J. Inorg. Chem.* 2004, v. 49, S45-S56.

[25] Belov, N. B. *Sketches on structural mineralogy*; Nedra: Moscow, Russia, 1976; 344 p.

[26] Yartys` V. A., Burnasheva V. V., Semenenko, K. N. *Uspekhi khimii (Rus).* 1983, v. 52, 529-562.

In: Neutron Diffraction
Editor: Xinzhe Jin

ISBN: 978-1-62808-725-3

Chapter 3

STUDY ON RESIDUAL STRAIN INDUCED BY THERMAL HYSTERESIS IN A COMPOSITE MATERIAL

Xinzhe Jin*

High Energy Accelerator Research Organization (KEK),
Ibaraki, Japan

ABSTRACT

A striking property of neutron beams is their high penetration depth, which has lead to their application to many composite materials. In particular, neutron diffraction is an important method for studying superconducting composite. In superconducting wires such as A15 and REBCO, it is known that the critical current has a dependence on the strain in a magnetic field. Since the superconducting wire is usually composed of three or more materials, residual strain is induced by the different coefficients of thermal expansion in the materials during the cooling process after the superconducting phase transition. From neutron diffraction measurements at room temperature, we previously reported that the residual strain in superconducting filaments in the wire is tensile and has an effect on the tensile strain dependence of the critical current. We also reported that we have effectively reduced the residual strain by using a thermal cycle method after the superconducting phase transition.

* E-mail: shinte@post.kek.jp; xinzhe.jin@riken.jp; kinnsinntetu@yahoo.co.jp (Present address: Riken, Kanagawa, Japan).

Studies on such thermal cycle are necessary to understand the mechanism of the residual strain of composite wire. In this paper, I present a theoretical analysis of the residual strain changes in a superconducting wire produced by thermal hysteresis based on neutron measurement result.

Keyword: Neutron diffraction, thermal hysteresis, multi-peak analysis, composite material

INTRODUCTION

Residual strain in a composite material such as a superconducting wire has been previously studied because of the important applications of such materials. Thermal changes induce residual strain in a composite material that depends on the Young's modulus and coefficient of thermal expansion (CTE) of each material [1-3].

In a superconducting wire, residual strain is induced in the cooling process following the superconducting phase transition. A restraint-complete thermal stress in a single material is caused by a strain-constant constraint. However, in a composite material lacking an externally applied constraint, the materials in the wire interact with each other via the stress induced by the differences in the thermal strain between each material in process of the temperature decreasing. This stress is caused by the strain change; therefore, the thermal stress is not fully constrained and hence residual strains are induced in the materials. To understand the strain generative mechanism, it is necessary to study the lattice structure behavior over the temperature profile involved.

In RHQ-Nb3Al wires having different matrix materials, we previously reported that there are different residual strains in the Nb3Al filaments at room temperature in the axial direction of the wire [1]. This difference has an effect on the mechanical properties of the wire. A higher reversible strain limit of the critical current is produced in a sample having a smaller tensile residual strain in the axial direction of the Nb3Al filaments. Therefore, reducing the tensile residual strain is related to improving the mechanical properties, such as the ultimate tensile strength.

The residual strain change induced by thermal hysteresis has been studied in Bi2223 superconducting tape by Ochiai et al. [4]. For Nb_3Al superconducting wire, we have been reported the effect of thermal hysteresis on the residual strain with experimental results [5]. Reducing the residual strain of the superconductor in a composite wire, is beneficial for improving their performance of proof strength and avoid degradation of the superconductor by the electromagnetic force produced during magnet operating.

Controlling residual strain induced in superconducting wire by thermal changes, is one of the best ways to reduce damage introduced during fabrication. Residual strain changes caused by thermal hysteresis has been observed by neutron diffraction, however, a theoretical analysis has been lacking. In this paper, models of the strain induced by thermal changes in the longitudinal direction of composite wire are examined and compared with neutron diffraction measurements.

THERMAL STRAIN MODELS FOR A COMPOSITE MATERIAL

For all the materials in a composite wire, the various strain distributions were assumed to have no hysteresis due to thermal changes in the range of elastic deformation. Since a superconducting wire usually also contains a material that exhibits plastic deformation such as Cu, strain hysteresis of superconductor is induced by the interaction with the inelastic behavior of the Cu in the composite wire. With respect to this interaction, two models that simply employed one elastic and one inelastic material were considered that differed in the order of temperature changes.

The first model used a decreasing-increasing temperature process with Cu as the inelastic material, as shown in Figure 1. During cooling, as shown in Figure 1 (a) and (b), the strain in superconductor is lower than that in the single material, due to the transmitting stress from the Cu layer with its large CTE. The lowest temperature reached was defined as T_{L1}. As a result, plastic deformation was induced in the Cu along tensile direction. This permanent strain affected the internal strain distribution in the superconductor layer during the temperature recovery process.

On increasing the temperature in the recovery process, the residual stress between Cu and superconductor decreases to produce a residual stress-free state at a temperature T_{L2} between T_{L1} and T_0, as shown in Figure 1 (c). The tensile deformation of Cu is effective in decreasing the equilibrium temperature of this residual stress-free state.

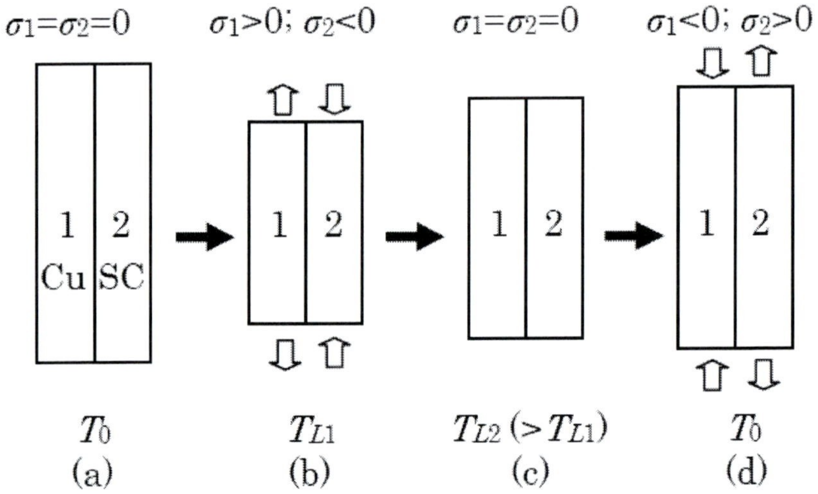

Figure 1. Strain model of the temperature decreasing-increasing process.

Above the temperature T_{L2}, compressive stress was induced in Cu due to it having a larger CTE than that of superconductor. When the temperature returns to its original T_0 value, residual stress was induced between the Cu and the superconductor by thermal hysteresis. The residual strain in superconductor is tensile. It can be seen that the decreasing-increasing temperature process has the effect of shifting the residual strain in the superconductor in the tensile direction.

The stress-strain behaviors for each material in the decreasing-increasing temperature process were shown in Figure 2 (a). Strain distribution in wire was changed by the permanent strain ε_p of Cu, and its corresponding stress σ_p is a redistribution component of the residual strain.

In the second model of an increasing-decreasing temperature regime, one difference from the first model is the direction of the permanent strain in Cu. With increasing temperature, the Cu acquires a compressive permanent strain by interaction with the superconductor in the elastic deformation region. When the composite was heated above the strain recovery temperature of Cu, the equilibrium stress-strain temperature increased to be equal to the recovery temperature, and is higher than the original temperature T_0. Therefore, the residual stress distribution induced between Cu and superconductor by thermal hysteresis in this case is different from that in the decreasing- increasing temperature process, when the temperature returns to its original value of T_0. A compressive residual strain was induced in the superconductor.

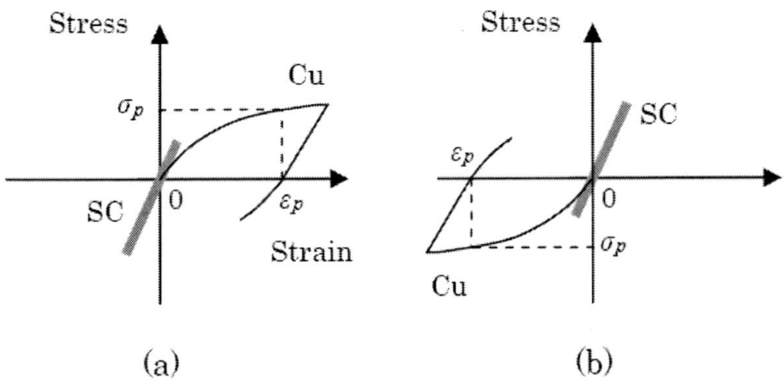

Figure 2. Stress-strain behavior of each material in composite wire during a thermal cycle (a) in temperature decreasing-increasing process, and (b) in temperature increasing-decreasing process, that remains below the recovery temperature of Cu). Permanent strain and its stress for Cu is defined as ε_p and σ_p, respectively.

Result, the two temperature change regimes have different effects on the residual strain in the superconductor. To reduce residual strain, the increasing-decreasing process is most effective.

The stress-strain behavior without thermal strain in the second process is shown in Figure 2 (b). The compressive stress σ_p was induced by the permanent strain ε_p in Cu as a component of compressive residual strain in the superconductor.

Neutron Diffraction Measurement Principles

To measure the residual strain in composite material, neutron beam is best used to obtain the lattice constants of internal material. The residual strain can be obtained from the lattice spacing d as follows:

$$\varepsilon_r = (d - d_0)/d_0$$

The lattice spacing d in the stress-free state is defined as d_0. Considering that permanent strain has almost no effect on lattice spacing, the residual strain obtained from the lattice spacing does not include the permanent strain. The total of these two strains is called the macro strain.

$$\varepsilon_m = \varepsilon_r + \varepsilon_p$$

Time-of-flight neutron diffraction is useful to measure the lattice spacing during thermal hysteresis. Because the statistics of this type of measurement depends on the beam flight time, results can be obtained with high accuracy if they are carried for a sufficient time at each temperature.

In these measurements, the sample is placed in a non-crystalline glass tube in a vacuum or in Ar gas. In the temperature control system, including a heater and thermometer, the latter should be attached near, or on, the sample. To obtain the actual values of the residual strain, the lattice spacing in the stress-free state is required, in, for example, a powder sample. If all that is required is to observe the temperature variation of the residual strain, that sample is not needed. Neutron diffraction measurement diagram for wire and powder are shown in Figure 3 (a) and (b), respectively. To observe the strain in the longitudinal direction of the wire sample, the angle between the longitudinal direction and the incident beam should be set to 45 degrees. This angle is not needed for a powder sample, but, the powder must to be enclosed in a sample holder in order to prevent its spreading.

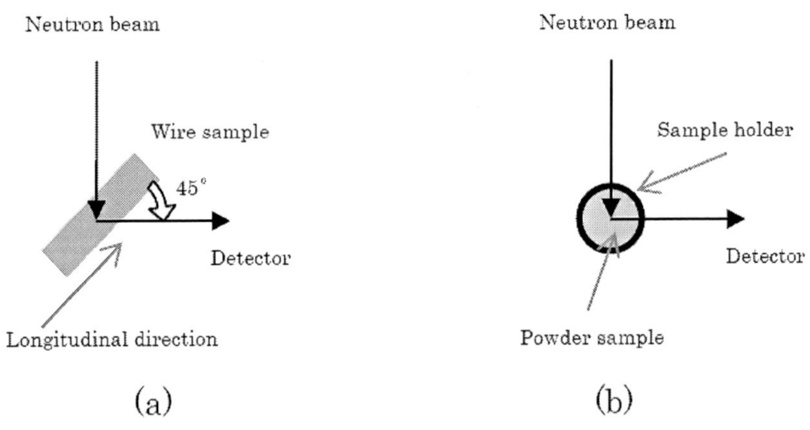

Figure 3. Neutron diffraction measurement diagram of time-of-flight method for (a) wire, and for (b) powder.

In the analysis method of the neutron diffraction patterns, we have previously used a multi-peak analysis [1], also have observed the residual strain change from a thermal cycle using a single-peak analysis [5]. The result of such a measurement was that the residual strain in the superconductor decreased markedly during the temperature increasing-decreasing process. This large change of the residual strain is due to the high thermal variation. For a small change on temperature, a multi-peak analysis may be useful to examine the small change in residual strain. In the multi-peak analysis, we have typically used least-square fitting for peak position. Alternatively, each peak can be found from a Rietveld analysis soft as multi-peak analysis. Considering calculation speed and background removal, the multi-peak analysis by using a Rietveld analysis soft is better to obtain the efficient and effective results.

The crystal texture of each material can be observed by comparing the neutron diffraction pattern of a sample with that in the strain free state. In this paper, the change of the texture was not considered due to its small variation. By comparing the peak intensity in each Miller index between the wire and powder, it can be seen whether all materials in the sample have texture or not. Due to the crystal texture, it is difficult to determine the lattice parameters for a material by using the peak intensities in the neutron diffraction patterns. For this reason, the value of each lattice spacing d is obtained by finding the peak position by a single peak analysis in the thermal cycle process in order to determine the lattice structure change. The mean strain can be estimated from the lattice spacing d, by

$$\bar{\varepsilon}_r = \sum_{i=1}^{n} \frac{d_i - d_{i0}}{d_{i0}},$$

where n is the number of the analysis peak.

Another approach to the fitting method in multi-peak analysis, is to obtain the residual strain by determining the lattice constant a by a least-square approach for a specific phase in the neutron diffraction pattern. The lattice constant a for a textured sample is different to that for a powder model in physical meaning. The strain-free powder can be regarded as having almost no texture effect on the diffraction pattern. Therefore, static analyses for peak intensity and peak position have been studied for powder by a Rietveld analysis. Since the diffraction elastic coefficients of all the peaks in a material have almost the same value in the strain-free state, also it is no texture, in powder; it can be modeled on a single crystal structure. However, in the superconducting wires, the crystal texture is usually induced in their fabrication process such as wire drawing. Also the wires have a residual strain in each material, therefore, the diffraction elastic coefficient has dependence on the peak in a phase. The result is that the wire then falls outside the definition of the Rietveld analysis model. Therefore, a model for a multi-peak analysis of a wire is to normalized a virtual lattice structure, that in a powder is approximated by an ideal lattice structure. The crystal lattice obtained by multi-peak analysis does not actually exist, but the strain measurement accuracy is improved by statistics by a virtual crystal lattice, as has been already experimentally verified [1]. In A15 superconducting wire, strong textures are induced in the wire drawing process as a result of the area reduction. In REBCO coated conductor, the crystal orientation was almost aligned to improve the critical current density. Considering the texture effect on the peak intensity, during the intensity fitting of each peak in a phase should be set to a free fitting parameter without using any function. It is different from Rietveld analysis for a powder. The crystal structure is deformed into various shapes by the anisotropic residual strain in the composite wire.

To compare with the wire sample, the powder sample can also be analyzed by multi-peak analysis. The peak intensities for the powder were also fitted independently in each other. Due to the different diffraction elastic coefficient for each peak, residual strains obtained from the single-peak analysis have a variation among the peaks; the results indicate that the residual strain depends on the hkl values of the peaks. Considering these hkl peak variations, the residual strains found by the multi-peak analysis method are weighted-average values which may express macroscopic residual strains. The statistics of the multi-peak analysis in a composite material will be useful for a lot of studies in engineering strain.

Residual Strain Effect on Superconducting Wire Application

In magnet applications of superconducting wire, the discussion usually focuses on the electromagnetic force. However, it is difficult to understand the actual stress distribution. To express the stress on the wire in a magnet, the simple tensile stress along the longitudinal direction of the wire is susceptible to experimental evaluation. When investigating the tensile strain dependence of the critical current, the behaviors of superconducting wires depend on the residual strain. The tensile strain at the peak of the critical current is increased with a decrease in the residual strain. Then the reversible strain is increased with an improvement in the bearing force. The residual strain of the superconductor in the wire has a compression or a tensile strain that has dependence on the CTE and Young's modulus of the matrix material. Using thermal hysteresis to reduce the residual strain is one of the best ways to improve wire performance for high field magnet.

CONCLUSION

The residual strain induced by thermal hysteresis in a composite material was studied with two models that employ either increasing-decreasing or decreasing-increasing temperature profiles. The residual stress distributions induced between Cu and superconductor by thermal hysteresis show difference between the two models, when the temperature returns to its original temperature. The residual strain in the superconductor increased to be tensile in temperature decreasing-increasing process, but was reduced to compression in the temperature increasing-decreasing process. The residual strain can be obtained from the lattice spacing using a neutron diffractometer. Since permanent strain is almost no effect on lattice spacing, the residual strain obtained from the lattice spacing has no relation to the permanent strain, while the macro strain is their sum. Multi-peak analysis methods were described that can be used for powder and wire samples, whether or not they have texture or residual strain.

REFERENCES

[1] Jin X, Nakamoto T, Ito T, Harjo S, Kikuchi A, Takeuchi T, Tsuchiya K, Hemmi T, Ogitsu T and Yamamoto A, 2012, *Supercond. Sci. Technol.* 25, 065021.

[2] Jin X, Nakamoto T, Harjo S, Ito T, Ogitsu T, Tsuchiya K, Yamamoto A, Kikuchi A, Takeuchi T and Hemmi T, 2012, J. *Alloys Compds*. 535, 124-128.

[3] Jin X, Oguro H, Nakamoto T, Awaji S, Yamamoto A, Tsuchiya K, Ogitsu T, Kikuchi A, Takeuchi T, Harjo S, and Ito T, 2012, *Cryogenics* 52, 805-809.

[4] Ochiai S, Rokkaku H, Morishita K, Shin J, Iwamoto S, Okuda H, Hojo M, Osamura K, Sato M, Otto A, Harley E, and Malozemoff A, 2007, *Supercond. Sci. Technol.* 20, 202-210.

[5] Jin X, Nakamoto T, Tsuchiya K, Ogitsu T, Yamamoto A, Kikuchi A, Takeuchi T, Harjo S, Ito T, and Tomota Y, 2013, *IEEE Trans. Appl. Supercond.* 23, 6000704.

In: Neutron Diffraction
Editor: Xinzhe Jin
ISBN: 978-1-62808-725-3

Chapter 4

THE NEUTRON DIFFRACTOMETERS FOR FATIGUE STUDY OF ENGINEERING METALS

E-Wen Huang[1*], Soo Yeol Lee[2] and Chung-Hao Chen[3]

[1]Department of Chemical & Materials Engineering and Center for Neutron Beam Applications, National Central University, Jhongli City, Taiwan (R.O.C.)

[2]Department of Materials Science & Engineering, Chungnam National University, Daejeon, South Korea

[3]Department of Electrical and Computer Engineering, Old Dominion University, Norfolk, VA, United States

ABSTRACT

Fatigue failure is one of the most practical engineering challenges. Hence, some of the engineering components sacrifice their service time and/or operation limits to fulfill the estimated safety factors. However, in terms of the efficiency and environmental perspectives, if there comes more mechanistic understanding of the fatigue behaviors, the potential of the engineering materials could be fully economically appreciated for extending their service potential. In fact, the fatigue behavior has been studied for decades. Various methodologies have been developed to

* E-mail: ewhuang@ncu.edu.tw.

describe the driving force of the fatigue behaviors in terms of the plasticity and the fatigue crack behavior. Most of the aforementioned understanding is based on the utilization of the electron microscopy-based experimental measurements and their theoretical calculations. Many phenomenological mechanisms can relate the bulk fatigue life to the microscopic mechanisms. However, the current understanding still faces some challenges, such as the sampling volume to correlate the lattice information and the bulk behavior. Recently, several international neutron facilities bring new class of engineering diffractometers. These new instruments, equipped with the load frames and furnace, enable the *in-situ* measurements to investigate the microstructural information to interpret the cyclic effects at different environmental temperatures. The advantages of these new devices provide high-penetration neutron to gauge greater amount of the sampling sizes for correlating the bulk fatigue behavior with the lattice-level information, which sheds light on a new chapter of the fatigue study. In this chapter, the fatigue behavior study via neutron diffractometer will be summarized.

Keywords: Neutron diffraction, Fatigue, Cyclic Loading, Crack Mapping

INTRODUCTION

Fatigue failure is one of the most frequent accidents for most of the engineering applications. The engineering society has developed both of the characterization methods [1] and well-defined theories [2] to investigate either the dislocation-based cyclic-loading activities [3] or crack-based fatigue [4] problems. However, there is still a gap to clarify the transition between the dislocation deformations to the fatigue crack failure because it is not easy to investigate the fatigue-induced micro voids [5]. Moreover, except the simulation [6], it is not trivial to capture the evolution of the fatigue-induced microstructure with nondestructive methods. Recently, several nondestructive experiments have demonstrated the potential to investigate the materials characteristics subjected to cyclic loadings, such as the nonlinear ultrasound [7], the neutron diffraction [8], the synchrotron radiation [9], and the imaging [10] measurements.

Among these complimentary methods, we will introduce why and how to apply the neutron diffraction together with the others to reveal the fatigue behavior of the metals subjected to the cyclically-loading damage.

Neutrons are neutral particles. They have no interaction, nor electrostatic with the electrons surrounding an atom.

Table. Selected practical fatigue experimental studies using neutron diffractometers

Neutron Instruments and website links (alphabetical order)	Title of the publication	Fatigue issue
G-5.2 diffractometer at the Laboratoire Léon Brillouin, France (http://www-llb.cea.fr/en/fr-en/spectros_p.php) TAS8 at Risø National Laboratory, Denmark (http://www.risoecampus.dtu.dk/DTU_Fysik/Projects_Neutrons.aspx)	Separation of measured fatigue crack stress fields in a metal matrix composite material [13]	Metal-matrix-composite fatigue crack
ISIS - ENGIN-X (http://www.isis.stfc.ac.uk/instruments/engin-x/engin-x2900.html)	Twinning-detwinning behavior during fatigue-crack propagation in a wrought magnesium alloy AZ31B [16]	Twinning-detwinning on fatigue crack
KOWARI-ANSTO (http://www.ansto.gov.au/research/bragg_institute/facilities/instruments/kowari)	Effect of residual stress on the integrity of a branch connection [11]	Welded branch connection
L3 spectrometer-CNBC (http://www.nrc-cnrc.gc.ca/eng/facilities/cnbc/spectrometers/l3.html)	On the generation of microstrains during the plastic deformation of Waspaloy [12]	Orientation dependence of the microstrain
LVR-15 (http://neutron.ujf.cas.cz/en/lwr15-description)	In-situ neutron diffraction during shape-memory behavior in Fe-Mn-Si-Cr [18]	Cyclic shape-memory behavior
NRSF2-HFIR (http://www.sns.gov/instruments/HFIR/HB2B/)	A study on fatigue crack growth behavior subjected to a single tensile overload: Part II. Transfer of stress concentration and its role in overload-induced transient crack growth [15]	Fatigue random-load effect
RSI-HANARO ((http://nsrc.jaea.go.jp/aonetned/rsi.pdf))	Three-Orthogonal-Direction Stress Mapping around a Fatigue-Crack Tip Using Neutron Diffraction [14]	Fatigue-crack mapping
SMARTS-LANSCE (http://lansce.lanl.gov/lujan/instruments/SMARTS/index.html)	Fatigue-induced reversible/irreversible structural-transformations in a Ni-based superalloy [17]	Low-cycle-fatigue and thermal responses
TAKUMI-JPARC (http://j-parc.jp/MatLife/en/instrumentation/ns_spec.html)	Development of neutron detector for engineering materials diffractometer at J-PARC [19]	Creep fatigue
VULCAN-ORNL (http://neutrons.ornl.gov/instruments/SNS/VULCAN/)	The development of grain-orientation-dependent-residual stressess in a cyclically deformed alloy [20]	Fatigue-induced texture

Thus, the neutrons can be scattered almost elastically off the nucleus of materials for the non-energy-exchange measurements. Moreover, the size of nucleus of an atom is much smaller than its atom. Hence, neutrons have a much smaller cross section than those of electrons or photons. Besides the effect of the physical sizes, the nuclear-effective radii do not change with the atomic number (Z) directly. Because of this, the atomic-scattering factor from neutrons does not follow the Z-number-dependent X-rays scattering. This unique cross-section characteristic allows the scientist to design experiments to facilitate the isotopes and so on to map the differences between atoms of similar atomic numbers. In terms of the levels of the perturbation during the measurements, neutron-diffraction experiments have much lower energies than that of typical X-ray radiation. The weak interaction between the neutrons and the measured subjects, coupled with the small cross section of the nuclei of the subjects, allows neutrons to penetrate much deeper than the rest of the aforementioned complementary measurements, such as electrons and X-rays. Hence, the neutrons do provide unique features for greater sampling underneath information without damaging the illuminated samples. This trend allows neutron diffraction to be used as a nondestructive tool on engineering parts and fragile samples. Since many records referring to fatigue and neutron are well documented [9], only few general cases are introduced here. The present status on fatigue study of engineering metals relating to neutron diffraction experiments is summarized in the following tables based on the Thomson Reuters data base of the Web of Knowledge.

As summarized above, using neutron diffractometers to solve the fatigue issues is already a practical method; some important pending problems in the fatigue study are introduced below first. The application of the neutron diffraction experiments to answer those challenges is referred in the following paragraphs. Two types of information with respect to fatigue study are separately introduced. The first category of information relates to the primary information directly and independently obtained from the neutron diffraction experiments. The secondary one is the indirect information concluded by using the neutron diffraction data like dislocation sub-structure for fatigue-damage mechanism study.

1. Important Issues of the Fatigue Damages

The fatigue behavior is highly related to the microstructure evolution subjected to cyclic loadings [21-22]. Fatemi and Yang [23] reviewed the fatigue-life models based on the crack growth concept, continuum damage mechanics, the linear and nonlinear damage rules; two-stage linearization approaches, and energy-based theory. Among the phenomenological and semi analytical theories, there are no clear boundaries due to the cumulated nature of the fatigue damages. Mughrabi [24] concludes that the dislocation activity is responsible for the degradation during cyclic loading. Guinier and Julien[5] addressed the importance of the microcracks for the fatigue-damage initiation. Above all, some key parameters are concluded for systematic understanding. They are the fatigue crack growth, the initiation of the fatigue crack, and the dislocation activities subjected to the cyclic loadings. However, the size differences of the microstructures between the dislocations [25] and the cracks [26] can cross several length scales. Hence, it is not easy to conclude a governing formula to describe such a multi-scale fatigue behavior. To bypass the direct mechanistic interpretation, many statistical-based approaches have also been applied to predict the fatigue life based on the probability [27]. Based on the aforementioned perspectives, this chapter introduces the different applications of the different neutron diffractometer for each perspective to study the various fatigue behaviors. The focus is on the structural materials, especially those metallic systems.

2. Fatigue-Crack Propagation Subjected to Loadings

When the fatigue damage on the engineering components forms the cracks at certain size, the dislocation microstructure cannot correlate the bulk fatigue behavior properly. Instead, fatigue-induced crack propagation becomes a dominant mechanism during the fatigue. Generally the fatigue crack propagation involves random or variable amplitudes of loads rather than constant-amplitude loadings [22, 28, 29]. Few sudden changes during the long constant cyclic loadings could dramatically influence the plastic zones and, hence, the distribution of the lattice-stress states near a fatigue crack can give rise to a significant crack-growth retardation and/or acceleration. The reality of the complicated cyclic loadings makes it difficult to quantify the fatigue-crack propagation and its associated life. Therefore, overload and underload effects are characterized to correlate the damage-tolerance design and lifetime-prediction methodology.

As a simplest case, the effects of a single tensile overload on fatigue-crack growth have been studied since the 1960s. Numerous experimental studies[29]-[31, 32] have shown crack-growth-retardation phenomena. There are also computer simulation studies [33, 34]. This sort of tensile-overload condition gives the beneficial effects, which improve the fatigue life. The application of a single tensile overload during fatigue-crack growth results in a large crack-growth retardation period, which significantly increases the fatigue lifetime. Elber [30] proposes that there is the plasticity-induced crack closure in the wake of a crack and effective-stress-intensity-factor range as a fatigue-crack-tip-driving force.

A crack-closure approach has played an important role in explaining many load-interaction effects on the fatigue-crack-growth behavior under variable-amplitude loading [35]. Nowadays, it is well accepted that the crack opening and closing loads is important to predict the crack-tip-driving force.

Most of the experimental crack-closure measurements are based on the analysis of the specimen compliance, i.e., displacement/load [30, 36-37]. Various possible mechanisms have been proposed to account for the overload-induced transient crack-growth phenomena, which include the crack-tip blunting [38-39], compressive residual stress ahead of the crack tip [40], crack branching [32], plasticity-induced crack closure [30, 41], and roughness-induced crack closure [32]. It has been argued that the crack-tip blunting by the overloading can persist even during post-overload crack growth, and lead to crack-propagation retardation [38]. The concept suggests that the blunted crack tip behaves like a notch with a less severe stress concentration than the originally sharp crack tip. Although crack-tip blunting does influence the post-overload crack-growth rate, it cannot account for the existence of delayed retardation and cannot quantitatively rationalize the experimentally-observed reductions in post-overload crack growth [32].

Recently, there are also other claims to conclude the fatigue behavior is not only related to the crack closure, but the fatigue-crack-tip-driving force [42, 43-45]. Sadananda et al. [44] report that the perturbation of the stresses ahead of the crack tip is the major factor for the overload retardation. A "unified approach", which the maximum stress-intensity factor, K_{max}, and the stress-intensity-factor range, ΔK (= $K_{max} - K_{min}$), are considered as the two driving forces required for fatigue-crack growth.

So far, the retardation micromechanism, fatigue-crack-tip-driving force, and the crack-closure phenomenon in a fatigue wake still an open question. The reason is the experimental difficulties to quantify strain/stress fields near a fatigue-crack tip under applied loads and to observe real-time fatigue-crack-tip deformation and failure phenomena during cyclic loadings.

2.1. Direct Neutron-diffraction Experimental Approaches to Investigate Fatigue Cracks

The determination of crack opening and closing stresses is important to understand the fatigue-crack-propagation behavior. Among several possible retardation mechanisms, the plasticity-induced crack closure has received considerable support in the fatigue community. However, other investigators have argued that plasticity always opens the crack rather than closing the crack [43, 46]. Sadananda et al. [44] reported that the residual stress ahead of the crack tip is a major factor for the retardation, rather than the crack closure behind the crack tip, and suggested a new "unified approach" in which both ΔK ($K_{max} - K_{min}$, the stress-intensity-factor range) and K_{max} are the fatigue-crack-tip-driving force. Before the introduction of how the neutron diffraction can answer the aforementioned fatigue-crack issues, the following is a summary of several conventional methods to investigate the crack behaviors subjected to the cyclic loadings.

A number of different experimental techniques, such as the compliance-based method [36, 47], moiré interferometry method [48], ultrasonic method [49], potential-drop technique [50], acoustic-emission technique [51], photoelasticity method [52], and synchrotron X-ray microtomography [53] are used to study the load at which the crack tip opens. Moreover, analytical models [54] have also been developed to predict the crack opening and closing stresses. Many researchers have simulated the plasticity-induced crack closure in the two-dimensional geometries using the finite-element analysis under plane-strain [34, 55] or plane-stress conditions [56]. The materials microstructure measured and simulated by these methods will be introduced to bring in the ideas of the fatigue cracks.

First, the direct-current-potential-drop (DCPD) technique is an experimental method to measure the crack closure. The crack-closure phenomena during each loading-unloading cycle can be observed by the changes in the electric potential. When a constant electric current conducts through the specimen, the associated crack-mouth potential is measured. Based on the fundamental potential-current relationship, the higher potential means the longer crack length due to an increased resistivity of the material. As the crack propagates, the resistivity of the material is increased, and, thus, the change of the crack-mouth potential occurs. By detecting the change of the potential drop, the corresponding crack length is calculated. When a fatigue crack closes, which yields an electric contact between the fracture surfaces, the crack-closing point can be revealed by the transition point on the curve of the applied load vs. potential subjected to fatigue cycles [51].

The other experimental crack-closure studies are mostly based on the analysis of the specimen compliance. An estimate of the crack-opening stress level can be derived from a record of the stress (load) versus crack-opening displacement (COD). The fatigue specimen has different linear and nonlinear behaviors. The first linear behavior exhibits the elastic response of the partially closed-crack face.

The nonlinear portion represents the gradual crack-tip opening from the surface to the interior. The subsequent beginning of another linear response reveals the load level which the crack tip is fully open. Generally, the final nonlinear region represents the loading phase where the plastic deformation ahead of the crack tip dominates. For example, Andersson et al. [57] examined the crack-closure with *in-situ* scanning-electron microscope (SEM). They found that, for the crack-closure measurements, the potential drop results similar to closure measurements determined from *in-situ* SEM observations, and concluded that the crack-opening point is reliable if crack closure is detected by potential measurements.

To sum it up, for a single tensile overload, Sadananda et al. [44] concluded that all factors that influence the plasticity at the crack tip will have direct or indirect influence on overload effects. The retardation is generally measured in terms of delayed cycles, N_d. The retardation effect depends on the overload ratio (OLR), the background ΔK, where a single tensile overload is applied, and the background R ratio. Retardation persists until the crack has propagated out of the perturbed plastic zone, a distance related to both the background plastic zone and the spectrum of the overload.

Certainly, there are still debates regarding to the crack-closure mechanisms [42, 43-45], which a variety of nondestructive-diffraction techniques are applied to map the microstructure around the crack tips. They are X-ray diffraction and tomography [45, 58], and neutron diffraction [59, 60] utilized to study the overload effects on fatigue-crack growth.

The above nondestructive-diffraction techniques can map the residual-stress fields in the vicinity of the crack tip on the overloaded samples, which the other instruments cannot. The orientation-dependent information, such as the transverse residual stresses can be spatially-resolved as a function of the distance from the crack tip.

2.2. Engineering Diffractometer to Map Fatigue Cracks

Unlike the other approaches to study the fatigue cracks, real-time fatigue experiments coupled with *in-situ* diffraction measurements focusing on a fatigue-crack tip can reveal stress/strain fields and plastic deformation of the crack-tip deformation and failure characteristics. Several synchrotron x-ray stations and neutron diffractometers have reported the fatigue-crack mapping results, which correlate the lattice-level micromechanism to the fatigue-crack-tip-driving force and the crack-closure phenomenon. The differences are majorly based on their instrumental spatial resolutions. With the smaller synchrotron coherent-focus size, the synchrotron x-ray results can offer finer mappings than the neutrons. For example, Croft et al. [45] map the lattice-strain evolutions in the vicinity of the crack tip with *in-situ* synchrotron X-ray diffraction. They found that the lattice-strain change ($\Delta\varepsilon_{yy}$) and maximum strain (ε_{yy}) are correlated to the driving force of crack tip using a unified approach (the combination form of ΔK & K_{max}). The aforementioned lattice strains can be obtained from the aforementioned Equation (1), where interplanar spacings (*d*-spacings) along the longitudinal, transverse, and normal directions can be determined from the Gaussian fitting of the diffraction peak, and where d_0 is the stress-free reference *d*-spacing, which was measured away from the crack tip. Moreover, Three residual stress components, σ_i ($i = x, y,$ and z, corresponding to longitudinal, transverse, and normal directions, respectively), can be calculated from the three strain components using the following equation:

$$\sigma_i = \frac{E}{1+\nu}\left[\varepsilon_i + \frac{\nu}{1-2\nu}\left(\varepsilon_x + \varepsilon_y + \varepsilon_z\right)\right] \tag{1}$$

where E is the Young's modulus and ν is the Poisson's ratio. Their results demonstrate that the lattice-strain levels, such as $\Delta\varepsilon_{yy}$ and ε_{yy}, can be used as indicators for the crack-tip stress intensities ΔK and K_{max}.

However, in some engineering cases, the higher penetration levels are required. For example, some metal systems at the sizes of the American Society for Testing and Materials (ASTM) standards are too thick for the synchrotron x-ray measurements. Hence, neutron diffraction can offer unparallel capability to investigate the underneath greater gauge volumes. For examples, the neutron-diffraction residual-stress mapping can be performed on the Residual Stress Instrument (RSI) at the High-flux Advanced Neutron

Application Reactor (HANARO), Korea Atomic Energy Research Institute (KAERI) and KOWARI and WOMBAT diffractometers of Australian Nuclear Science and Technology Organisation (ANSTO). More information and publication can be found on the websites summarized at the link of the Asia-Oceania Network on Neutron Engineering Diffraction (AONET) (http://nsrc.jaea.go.jp/aonetned/index.html).

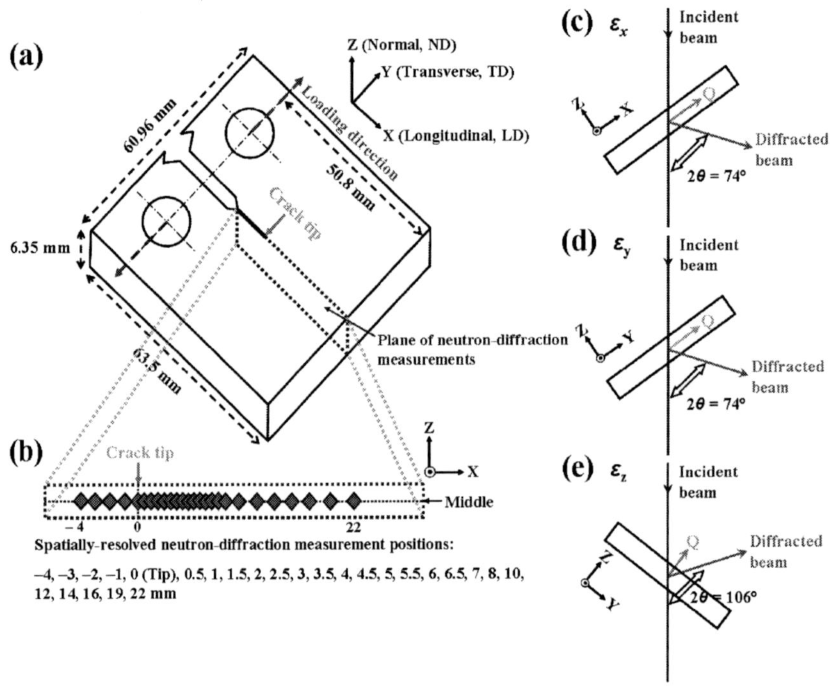

Figure 1. (a) The geometry of a compact-tension specimen; (b) spatially resolved neutron-diffraction measurement positions along the direction of crack propagation (x); Schematic of diffraction geometry for the residual-stress mapping showing the scattering vector (Q) parallel to the coordinate (c) x: longitudinal strain (ε_x) component; (d) y: transverse strain (ε_y) component; and (e) z: normal strain (ε_z) component.

The *in-situ* neutron experiments can be applied to directly determine the lattice strains and stresses near the crack tip.

To map the microstructure near a fatigue crack, the compact-tension (CT) specimen geometry can be applied because we can correlate the diffraction results with the bulk fatigue tests. A schematic view of the diffraction geometry is shown in Figure 1.

The specimen will be aligned relative to the incident neutron beam. The longitudinal (ε_x), the transverse (ε_y), and the normal (ε_z) strain components, as shown in Figs. 2(c), 2(d) and 2(e), can be resolved, respectively.

A series of the fatigue-crack investigation facilitating neutron diffraction experiments can be found in Lee et al.'s latest work [15, 61].

Above all, the *in-situ* measurement techniques enable the residual strain/stress mapping in the bulk sample as a function of the distance from the crack tip simultaneously.

The direct measurements of residual and internal strain variations near the crack tip under various loading conditions are of importance to further development of modeling work and to the advancement of a fundamental understanding of the crack-tip deformation and fracture behavior.

3. Cyclically-Induced Microstructure Evolution and the Associated Measurements

At the early stages of the cyclic loadings, there are already undergoing microstructure evolutions developments. The indirect information concluded by using the neutron diffraction data like dislocation sub-structure, fatigue-damage mechanism and so on will be reviewed. Before the description of the neutron experiments, other complimentary experimental methods are summarized.

For example, some resonant ultrasound spectroscopy (RUS) [62], evaluating elastic constants in materials [63], demonstrate that there do exist fatigue damages changing the elastic resonance frequencies of a sample using small perturbations created by ultrasonic frequencies[7]. The elastic tensor of a solid can be correlated to the damages [64].

Similarly, inelastic effects can be characterized using RUS such as microcracks in ceramics [65] and internal friction [66].

Furthermore, other experiments can directly quantify the lattice-level microstructure changes subjected to the cyclic loadings, such as the electron microscopy [21-22].

The microstructure evolutions subjected to the cyclic loadings can be texture evolution and/or dislocation changes at the lattice levels. If an ASTM-standard size is required to compare with the bulk fatigue performance, neutron diffraction can gauge most of the engineering alloys for fatigue study.

3.1. The Neutron Diffractometers for the Cyclically-Induced Structural Evolution

The spallation neutron sources have the advantages to illuminate multiple diffractions for texture-sensitive study as well as the nano-precipitate diffraction at the higher d-spacing. Some of the most advanced neutron diffractometers facilitating the spallation neutron source are introduced below as an example to demonstrate the neutron-measurement methods. They are VULCAN at the Spallation Neutron Source (SNS) of the Oak Ridge National Laboratory (ORNL) and TAKUMI of the Japan Proton Accelerator Research Complex (JPARC) as shown in the Figure 2.

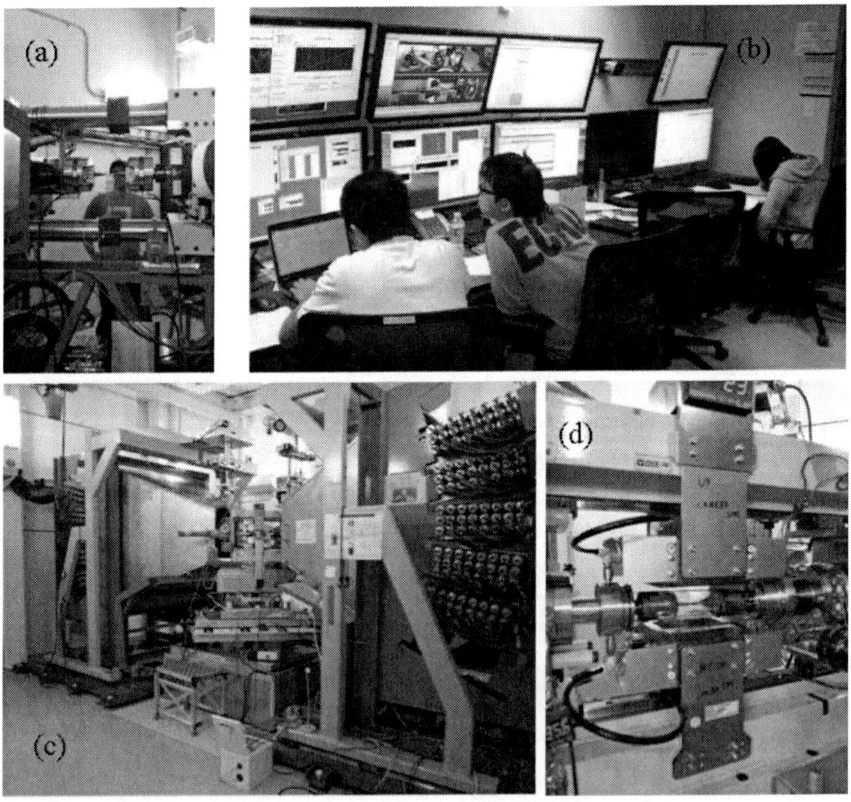

Figure 2. Spallation-neutron-source diffractometers: (a) VULCAN instrument; (b) VULCAN-remote control room; (c) TAKUMI at JPARC, and (d) the furnace and load frame of the TAKUMI.

VULCAN has equipments for various *in-situ* experimental environments, which make itself a powerful tool to study fatigue behavior of the structural materials. The main advantage of the *in-situ* neutron environment of VULCAN is to investigate the dynamics of lattice-evolution subjected to different levels of fatigue cycles. The hydraulic systems, Figure 2(a), and the remote control, Figure 2(b) are available. Moreover, newly-developed VULCAN software, VDRIVE, can reduce the data right after each end of every test. A continuous run can be conducted, taking advantage of VULCAN's event-based data acquisition, a unique feature at the SNS, where each neutron carries a time stamp. In this case, the diffraction data will be collected continuously and binned latter according to the desired time scale. Neutron diffraction data over several time frames can be averaged to yield the structure evolution within a load cycle. Similarly, TAKUMI can also create various *in-situ* environments, not limited to the fatigue research as well. TAKUMI's orientation-dependent-data-acquisition function can couple various detectors, Figure 2(c), and the software of the TAKUMI to resolve the preferred-texture-dependent diffractions.

3.2. Fatigue-Induced Microstructure Measurements Using Neutron Diffraction

Besides VULCAN and TAKUMI, Spectrometer for MAterials Research at Temperature and Stress (SMARTS) of the Los Alamos Neutron Science Center (LANSCE) and ENGIN-X of the ISIS at the Rutherford Appleton Laboratory will be recalled below for introducing the examples of the fatigue-induced-microstructure measurements. First, since the neutron can penetrate most of the metallic systems, a dog-shape specimen following the ASTM specifications can be applied. The cylindrical samples, 19.05 mm in length with 6.35 mm in diameter, can be mounted in a hydraulic load frame, with two perpendicular detector banks. The two detector banks simultaneously collect the diffraction from polycrystalline grains parallel (the loading direction, LD) and perpendicular (the transverse direction, TD) to the applied-load direction, respectively. The gauge volume can be as much as about 120 mm^3 for statistically-sufficient measurements to correlate the microstructure to bulk properties. The load frames of the aforementioned neutron stations have similar capability as a Materials Test System (MTS) machine, model 810. To analyze the diffraction profiles measured, several statistic functions for diffraction-profile fitting [67] can be selected with the General Structure

Analysis Software (GSAS) [17]. The refined neutron diffraction results can yield the lattice-strain, normalized intensity, and peak-width evolutions, respectively. The calculations of the lattice strain (ε) can be referred to the Equation (2) where d is the d-spacing and the d_0 is the reference d-spacing.

$$\varepsilon = (d-d_0)/d_0 \tag{2}$$

3.3. Lattice-Asymmetry, Texture, Nano precipitate, Dislocation Evolution Subjected to the Fatigue Damage

How would the neutron diffraction experiments characterize the micro-fatigue-crack initiation? An indirect fatigued-induced lattice-asymmetry phenomena has been proposed recently to demonstrate the possibility to apply neutron diffraction measurements to identify the onset fatigue cycles of the cracks [17]. The experiments show that, both at 300K and at an elevated-temperature experiment of 400 K, there does exist such a fatigue-induced lattice-asymmetry consequences [68]. The measurements were conducted on the Spectrometer for Materials Research at Temperature and Stress (SMARTS) of the Manuel Lujan Jr. Neutron Scattering Center, Los Alamos Neutron Science Center (LANSCE). The application of the Spallation Neutron Source neutron-diffraction experiments for mechanical-testing experiments can be referred to several articles [20, 69]. The experiments also coupled with the attached high-resolution thermocouples on the specimen to compare the bulk temperature evolution to the lattice-asymmetry parameters as a function of fatigue cycles. A fatigue deformation-induced lattice-strain asymmetry evolution subjected to fatigue damage have been defined as shown in Equation (3) [17]. The lattice-strain asymmetry is quantified as a grain-orientation-dependent transverse/loading parameter, defined as:

$$P \equiv P_{Tension/compression} = \frac{-\varepsilon_{transverse}}{\varepsilon_{loading}} \tag{3}$$

The asymmetry parameter shows irreversible structural transformations during fatigue. This stain-asymmetry evolution reveals the irreversible plastic deformation subjected to the fatigue. The tension/compression asymmetry in loading/transverse directions is clearly evident in the behavior of an asymmetry parameter P. The irreversible fatigue phenomena might relate to the formation of the microcracks. Cyclic-loading effects on the reported nickel-based superalloy, investigated by the aforementioned *in-situ* neutron-diffraction measurements, also show the temperature evolution. The calculated thermoelastic responses, based on neutron data, are compared with the measured bulk temperature evolution. Two transitions in the temperature-evolution are observed. The first transition is associated with the fatigue hardening/softening transformation. The second transition, observed at larger number of fatigue cycles, is coincidental with the asymmetry P, combining with the modeling [68], the results can be reasoned to the transitions of the fatigue damages. The characterizing ability on both the lattice-level information and the bulk performance can only be illuminated by the *in-situ* neutron experiments.

Another important feature of the spallation neutron source is that the spectrometer can collect multiple *hkl* diffraction peaks simultaneously. It is important for some engineering alloys with more complicated symmetry, such as the magnesium alloys. Especially, the plastic-deformation behavior of the alloys in the structure of hcp system is not limited to the simple slip activities, but also twinning. Coupled with the arrangements of the multiple detectors, such as the TAKUMI's shown in Figure 1(c), the measured peak-intensity evolution can demonstrate the texture-reorientation evolution as fatigue progresses for these samples. The normalized intensity of each *hkl* plane is to divide the intensity ($I^{fatigued}$) subjected to the selected fatigue cycles by the reference intensity (I^0) before the deformation.

For example, Wu et al.'s [70] SMARTS results show that there are texture-reorientation arrangements of the Mg alloys subjected to the cyclic loading. In Wu *et al.*'s [70] results, the twinning and detwinning phenomenon subjected to fatigue is revealed, which was unavailable without the *in-situ* neutron experiments collecting the orthogonal diffraction simultaneously.

Besides the light-weight Mg alloys, another important type of the engineering metallic system is the precipitate-strengthened alloy. However, due to the complicated phases, how to correlate the each response of various phases to the bulk fatigue behavior becomes a challenge. An *in-situ* neutron diffraction measurement conducted at the ENGIN-X of ISIS show that the wide-range of spallation neutron spectrum can illuminate different phases simultaneously and those diffraction peaks at the higher d-spacings [71].

A nickel-based sample containing thermodynamically-stabilized $Ni_2(Cr,Mo)$-phase [72] precipitate was used in this study [73]. After the aging processing, the constituent atoms of the precipitates gradually establish into nano sizes [74]. Since the precipitates significantly changes their mechanical behavior [75], the investigated samples were annealed to dissolve most precipitates into the face-centered-cubic (FCC) matrix structure. The specimen before fatigue is practically texture free [76]. Room-temperature *in-situ* neutron-diffraction measurements on the fatigue experiments were reported for the alloys without_ENREF_15 and with the nano precipitates for comparison. The results provide new insight into the microstructure, subjected to fatigue damage, by comparing the ensemble-average neutron-diffraction results of two phases. Understanding how microstructures of the metal matrix and the strengthening precipitates deform under cyclic.

Finally, the characterization of the dislocation evolution, subjected to the fatigue damage, quantified by the neutron is introduced herein. The peak-width information can be correlated with the dislocations. The diffraction data can be refined based on the fatigue-induced microstructure. For example, if there is only randomly-distributed dislocation developed after the cyclic loading, a Gaussian broadening component can be assumed for a full-width half-maximum (FWHM) of the diffraction profile. The profile shape function used in profile fitting programs is shown in Equation (4) [77].

$$G = I_0 \exp\left(-\ln 2 \left(\frac{2\Theta - 2\Theta_0}{w} \right)^2 \right) \tag{4}$$

Here is peak intensity; w is half FWHM, and is the peak position.

If the microscopy results of the fatigued sample show that there are more than randomly-distributed dislocations, but also cell structures, certainly different peak-profile fitting models can be applied accordingly. For example, Mughrabi's composite model [3] suggest that edge dislocations group within walls and screw dislocations remain randomly distributed in the cell interior. Huang et al.'s research relates the peak broadening to the fatigue-induced microstructure evolution. The mechanical-induced energy stored in the deformed specimen has been reported to generally derive from the randomly-distributed dislocations, the cellular-dislocation structure, and the heterogeneous elastic-energy contribution of the mean stresses. As a result, changes of these microstructures are assumed to result in most of the observed peak broadening. For example, the different microstructure developments subjected to the tensile and fatigue deformation can be distinguished, respectively [78]. Moreover, another fatigued-induced cyclic hardening and softening feature can be quantified by the *in-situ* neutron diffraction measurements. The dislocation rearrangements, such as increase and annihilation, within a fatigue cycle can be resolved [78].

CONCLUSION

Above all, in this chapter, the fatigue behavior is summarized. Several conventional experiments are introduced and the complimentary neutron-diffraction experiments are reviewed. Based on the different perspectives of the fatigue-damage mechanisms, different neutron-diffraction measurements could be applied.

The neutron-diffraction mapping techniques are addressed for the readers to compare with the fatigue-crack theories and complimentary methods. For the earlier fatigue stages, the lattice-level microstructure, the lattice-strain and the dislocation-parameter evolution upon the cyclic loading can be correlated with the macroscopic-bulk-thermomechanical behavior by in-situ experiments. The crystallographic-slip and the associated dislocation activities during the plastic deformation can be revealed by the *in-situ* neutron-diffraction results. By refining the diffraction profiles with general structure analysis software, the peak components enable the investigation of dislocations and the average distance between patterned-dislocation structures, respectively. The microscopic ensemble-averaged neutron results and the bulk-property evolution upon cyclic loading show the non-monotonic development of the fatigue-induced dislocation sub-structures, texture reorientation, and the lattice asymmetry.

ACKNOWLEDGMENTS

The authors would like to take this opportunity to thank the many neutron beamline scientists, the neutron society, and the neutron programs of the National Synchrotron Radiation Research Center (NSRRC) who have helped and shared very valuable experiences over the years. EWH appreciates the support from National Science Council Programs NSC-101-2221-E-008-039-MY3 and NSC-102-3113-P-007-014. SYL was supported by the National Research Foundation of Korea (NRF) grant funded by the Korea government (MEST) (No. 2012M2B2A4029572).

REFERENCES

[1] D. Kuhlmann-Wilsdorf and C. Laird, "Dislocation behavior in fatigue," *Materials Science and Engineering* 27 (2), 137-156 (1977).

[2] Kiran Solanki, S. R. Daniewicz, and J. C. Newman Jr, "Finite element analysis of plasticity-induced fatigue crack closure: an overview," Engineering Fracture Mechanics 71 (2), 149-171 (2004); K. Sadananda and A. K. Vasudevan, "Fatigue crack growth mechanisms in steels," *International Journal of Fatigue* 25 (9–11), 899-914 (2003).

[3] H. Mughrabi, "Cyclic Slip Irreversibilities and the Evolution of Fatigue Damage," *Metall Mater. Trans. B* 40 (4), 431-453 (2009).

[4] R. O. Ritchie, "Mechanisms of fatigue-crack propagation in ductile and brittle solids," *Int. J. Fract.* 100 (1), 55-83 (1999).

[5] Guinier and R. Julien, *The Solid State from Superconductors to Superalloys*. (Oxford University Press, 1989).

[6] E. A. Repetto and M. Ortiz, "A micromechanical model of cyclic deformation and fatigue-crack nucleation in fcc single crystals," *Acta. Materialia* 45 (6), 2577-2595 (1997).

[7] Anish Kumar, Raghavendra R. Adharapurapu, J. Wayne Jones, and Tresa M. Pollock, "In situ damage assessment in a cast magnesium alloy during very high cycle fatigue," *Scr. Mater.* 64 (1), 65-68 (2011).

[8] Yan-Dong Wang, Hongbo Tian, Alexandru D. Stoica, Xun-Li Wang, Peter K. Liaw, and James W. Richardson, "The development of grain-orientation-dependent residual stressess in a cyclically deformed alloy," *Nat Mater* 2 (2), 101-106 (2003).

[9] P. J. Withers and Hkdh Bhadeshia, "Overview - Residual stress part 1 - Measurement techniques," *Mater. Sci. Technol.* 17 (4), 355-365 (2001).

[10] L. Liu, N. S. Husseini, C. J. Torbet, W. K. Lee, R. Clarke, J. W. Jones, and T. M. Pollock, "In situ synchrotron X-ray imaging of high-cycle fatigue crack propagation in single-crystal nickel-base alloys," *Acta. Materialia* 59 (13), 5103-5115 (2011).

[11] M. Law, O. Kirstein, and V. Luzin, "Effect of residual stress on the integrity of a branch connection," *Int. J. Pressure Vessels* Pip. 96-97, 24-29 (2012).

[12] H. J. Stone, T. M. Holden, and R. C. Reed, "On the generation of microstrains during the plastic deformation of Waspaloy," *Acta. Materialia* 47 (17), 4435-4448 (1999).

[13] M. E. Fitzpatrick, M. T. Hutchings, and P. J. Withers, "Separation of measured fatigue crack stress fields in a metal matrix composite material," *Acta. Materialia* 47 (2), 585-593 (1999).

[14] E. W. Huang, S. Y. Lee, W. Woo, and K. W. Lee, "Three-Orthogonal-Direction Stress Mapping around a Fatigue-Crack Tip Using Neutron Diffraction," Metallurgical and Materials Transactions a-Physical *Metallurgy and Materials Science* 43A (8), 2785-2791 (2012).

[15] S. Y. Lee, H. Choo, P. K. Liaw, K. An, and C. R. Hubbard, "A study on fatigue crack growth behavior subjected to a single tensile overload: Part II. Transfer of stress concentration and its role in overload-induced transient crack growth," *Acta. Materialia* 59 (2), 495-502 (2011).

[16] W. Wu, S. Y. Lee, A. M. Paradowska, Y. F. Gao, and P. K. Liaw, "Twinning-detwinning behavior during fatigue-crack propagation in a wrought magnesium alloy AZ31B," Materials Science and Engineering a-Structural *Materials Properties Microstructure and Processing* 556, 278-286 (2012).

[17] E. W. Huang, R. I. Barabash, B. Clausen, Y. L. Liu, J. J. Kai, G. E. Ice, K. P. Woods, and P. K. Liaw, "Fatigue-induced reversible/irreversible structural-transformations in a Ni-based superalloy," International *Journal of Plasticity* 26 (8), 1124-1137 (2010).

[18] Y. Tomota, S. Harjo, P. Lukas, D. Neov, and P. Sittner, "In-situ neutron diffraction during shape-memory behavior in Fe-Mn-Si-Cr," *JOM-J. Miner. Met. Mater. Soc.* 52 (10), 32-34 (2000).

[19] K. Sakasai, T. Nakamura, M. Katagiri, K. Soyama, A. Birumachi, S. Satoh, N. Rhodes, and E. Schooneveld, "Development of neutron detector for engineering materials diffractometer at J-PARC," *Nucl. Instrum. Methods Phys. Res. Sect. A-Accel. Spectrom. Dect. Assoc. Equip.* 600 (1), 157-160 (2009).

[20] Y. D. Wang, H. B. Tian, A. D. Stoica, X. L. Wang, P. K. Liaw, and J. W. Richardson, "The development of grain-orientation-dependent-residual stressess in a cyclically deformed alloy," *Nature Materials* 2 (2), 101-106 (2003).

[21] D. Kuhlmann-Wilsdorf, "The theory of dislocation-based crystal plasticity," *Philos Mag A* 79 (4), 955-1008 (1999).

[22] K. Vasudevan and K. Sadananda, "Classification of Fatigue-Crack Growth-Behavior," *Metallurgical and Materials Transactions A-Physical Metallurgy and Materials Science* 26 (5), 1221-1234 (1995).

[23] Fatemi and L. Yang, "Cumulative fatigue damage and life prediction theories: a survey of the state of the art for homogeneous materials," *International Journal of Fatigue* 20 (1), 9-34 (1998).

[24] H. Mughrabi, "Cyclic Slip Irreversibilities and the Evolution of Fatigue Damage," Metallurgical and Materials Transactions A-Physical *Metallurgy and Materials Science* 40A (6), 1257-1279 (2009).

[25] M. A. Meyers, A. Mishra, and D. J. Benson, "Mechanical properties of nanocrystalline materials," *Progress in Materials Science* 51 (4), 427-556 (2006).

[26] R. Zoughi and S. Kharkovsky, "Microwave and millimetre wave sensors for crack detection," *Fatigue & Fracture of Engineering Materials & Structures* 31 (8), 695-713 (2008).

[27] D. Gary Harlow, Peter K. Liaw, William H. Peter, Gongyao Wang, and Raymond A. Buchanan, "An approach to modeling the S–N behavior of bulk-metallic glasses," Acta Materialia 56 (13), 3306-3311 (2008); M. A. Hemphill, T. Yuan, G. Y. Wang, J. W. Yeh, C. W. Tsai, A. Chuang, and P. K. Liaw, "Fatigue behavior of Al0.5CoCrCuFeNi high entropy alloys," *Acta Materialia* 60 (16), 5723-5734 (2012).

[28] M. D. Sangid, H. J. Maier, and H. Sehitoglu, "A physically based fatigue model for prediction of crack initiation from persistent slip bands in polycrystals," Acta Materialia 59 (1), 328-341 (2011); H. L. Huang, N. J. Ho, and W. B. Lin, "The study of dislocation structures at fatigue crack tips in polycrystalline copper under various crack propagation rates at stage II: crack propagation," Materials Science and Engineering a-Structural Materials Properties Microstructure and Processing 279 (1-2), 261-265 (2000); H. L. Huang and N. J. Ho, "The study of fatigue in polycrystalline copper under various strain amplitude at stage I: crack initiation and propagation," Materials Science and Engineering a-Structural Materials Properties Microstructure and Processing 293 (1-2), 7-14 (2000); T. Luoh and C. P. Chang, "Dislocation-free zones in fatigued copper polycrystals," *Acta Materialia* 44 (7), 2683-2695 (1996).

[29] L. P. Borrego, J. M. Ferreira, J. M. P. da Cruz, and J. M. Costa, "Evaluation of overload effects on fatigue crack growth and closure," *Engineering Fracture Mechanics* 70 (11), 1379-1397 (2003).

[30] Elber W, *ASTM STP* 486, 230-242 (1971).

[31] O. E. Wheeler, "Spectrum loading and crack growth," ASME 94, 181-186 (1972); R.E Jones, *Engineering Fracture Mechanics* (1973); D. Gan and J. Weertman, "Crack Closure and Crack-Propagation Rates in 7050 Aluminum," Engineering Fracture Mechanics 15 (1-2), 87-& (1981);

Newman JC, "A crack-closure model for predicting fatigue crack growth under aircraft spectrum loading," ASTM STP 748, 53-84 (1981); S. Suresh, "Crack Growth-Retardation Due to Micro-Roughness - A Mechanism for Overload Effects In Fatigue," Scripta Metallurgica 16 (8), 995-999 (1982); C. M. Wardclose, A. F. Blom, and R. O. Ritchie, "Mechanisms Associated With Transient Fatigue Crack-Growth Under Variable-Amplitude Loading - An Experimental And Numerical Study," Engineering Fracture Mechanics 32 (4), 613-638 (1989); C. S. Shin and S. H. Hsu, "On The Mechanisms and Behavior of Overload Retardation In Aisi-304 Stainless-Steel," International Journal of Fatigue 15 (3), 181-192 (1993); J. D. Dougherty, T. S. Srivatsan, and J. Padovan, "Fatigue crack propagation and closure behavior of modified 1070 steel: Experimental results," Engineering Fracture Mechanics 56 (2), 167-187 (1997); C. Makabe, A. Purnowidodo, and A. J. McEvily, "Effects of surface deformation and crack closure on fatigue crack propagation after overloading and underloading," International Journal of Fatigue 26 (12), 1341-1348 (2004); C. Bichler and R. Pippan, "Effect of single overloads in ductile metals: A reconsideration," Engineering Fracture Mechanics 74 (8), 1344-1359 (2007); J. Codrington and A. Kotousov, "Crack growth retardation following the application of an overload cycle using a strip-yield model," *Engineering Fracture Mechanics* 76 (11), 1667-1682 (2009).

[32] S. Suresh, "Micromechanisms of Fatigue Crack-Growth Retardation Following Overloads," *Engineering Fracture Mechanics* 18 (3), 577-593 (1983).

[33] X. Zhang, A. S. L. Chan, and G. A. O. Davies, "Numerical-Simulation of Fatigue Crack-Growth Under Complex Loading Sequences," Engineering Fracture Mechanics 42 (2), 305-321 (1992); S. Roychowdhury and R. H. Dodds, "Effects of an overload event on crack closure in 3-D small-scale yielding: finite element studies," Fatigue & Fracture of Engineering Materials & Structures 28 (10), 891-907 (2005); K. D. Singh, K. H. Khor, and I. Sinclair, "Roughness- and plasticity-induced fatigue crack closure under single overloads: Finite element modelling," *Acta. Materialia* 54 (17), 4393-4403 (2006).

[34] S. Pommier, "Plane strain crack closure and cyclic hardening," *Engineering Fracture Mechanics* 69 (1), 25-44 (2002).

[35] Arkema WJ Schijve J, " Report VTH-217," (1976); Schijve J, *ASTM STP* 982, 5-34 (1988).

[36] P. K. Liaw, T. R. Leax, R. S. Williams, and M. G. Peck, "Influence of Oxide-Induced Crack Closure On Near-Threshold Fatigue Crack-Growth Behavior," *Acta. Metallurgica* 30 (12), 2071-2078 (1982).

[37] K. K. Brahma, P. K. Dash, and B. Dattaguru, "Observation of Crack Closure Using A Crack Mouth Opening Displacement Gauge," International Journal of Fatigue 11 (1), 37-41 (1989); W. Yisheng and J. Schijve, "Fatigue-Crack Closure Measurements on 2024-T3 Sheet Specimens," *Fatigue & Fracture of Engineering Materials & Structures* 18 (9), 917-921 (1995).

[38] Christensen RH, *Metal Fatigue*. (McGraw-Hill, New Nork, 1959).

[39] Bathias and M. Vancon, "Mechanisms of Overload Effect on Fatigue Crack-Propagation in Aluminum-Alloys," *Engineering Fracture Mechanics* 10 (2), 409-& (1978).

[40] Broek J. Schijve "Crack Propagation Based on a Gust Spectrum with Variable-Amplitude Loading " Aircraft Engineering 34 314-316 (1962); K. Sadananda and D. N. V. Ramaswamy, "Role of crack tip plasticity in fatigue crack growth," *Philosophical Magazine a-Physics of Condensed Matter Structure Defects and Mechanical Properties* 81 (5), 1283-1303 (2001).

[41] Roberts R Trebules VW Jr., Hetzberg RW, , ASTM STP 536, 115-146 (1973); A. P. Reynolds, "Constant Amplitude and Post-Overload Fatigue Crack-Growth Behavior in PM Aluminum-Alloy AA-8009," *Fatigue & Fracture of Engineering Materials & Structures* 15 (6), 551-562 (1992).

[42] K. Vasudevan, K. Sadananda, and N. Louat, "Reconsideration of Fatigue Crack Closure," Scripta Metallurgica Et Materialia 27 (11), 1673-1678 (1992); A. K. Vasudeven, K. Sadananda, and N. Louat, "A Review of Crack Closure, Fatigue-Crack Threshold and Related Phenomena," Materials Science and Engineering a-Structural Materials Properties Microstructure and Processing 188 (1-2), 1-22 (1994); J. Zhang, X. D. He, and S. Y. Du, "Analyses of the fatigue crack propagation process and stress ratio effects using the two parameter method," International Journal of Fatigue 27 (10-12), 1314-1318 (2005); C. Vallellano, J. Vazquez, A. Navarro, and J. Dominguez, "A micromechanical model for small fatigue crack growth: an approach based on two threshold conditions," *Fatigue & Fracture of Engineering Materials & Structures* 32 (6), 515-524 (2009).

[43] N. Louat, K. Sadananda, M. Duesbery, and A. K. Vasudevan, "A Theoretical Evaluation of Crack Closure," *Metallurgical Transactions a-Physical Metallurgy and Materials Science* 24 (10), 2225-2232 (1993).

[44] K. Sadananda, A. K. Vasudevan, R. L. Holtz, and E. U. Lee, "Analysis of overload effects and related phenomena," *International Journal of Fatigue* 21, S233-S246 (1999).

[45] M. C. Croft, N. M. Jisrawi, Z. Zhong, R. L. Holtz, K. Sadananda, J. R. Skaritka, and T. Tsakalakos, "Fatigue history and in-situ loading studies of the overload effect using high resolution X-ray strain profiling," *International Journal of Fatigue* 29 (9-11), 1726-1736 (2007).

[46] K. Sadananda and A. K. Vasudevan, ""Discussion of error in the analysis of the wake dislocation problem" - Reply," *Metallurgical and Materials Transactions a-Physical Metallurgy and Materials Science* 29 (4), 1359-1360 (1998).

[47] R. Sunder, "System for Automated Fatigue Crack-Growth Testing Under Random Loading," International Journal of Fatigue 7 (1), 3-12 (1985); Kirk MT Roberson LJ, ASTM STP 982, 230-246 (1988); K. Donald and P. C. Paris, "An evaluation of Delta K-eff estimation procedures on 6061-T6 and 2024-T3 aluminum alloys," *International Journal of Fatigue* 21, S47-S57 (1999).

[48] T. G. F. Gray and P. M. Mackenzie, "Fatigue Crack Closure Investigation Using Moire Interferometry," International Journal of Fatigue 12 (5), 417-423 (1990); L. J. Fellows and D. Nowell, "Crack closure measurements using Moires interferometry with photoresist gratings," *International Journal of Fatigue* 26 (10), 1075-1082 (2004).

[49] S. I. Rokhlin and J. Y. Kim, "In situ ultrasonic measurement of crack closure," *International Journal of Fatigue* 25 (1), 51-58 (2003).

[50] T. T. Shih and R. P. Wei, Engineering Fracture Mechanics 6, 19-32 (1974); V. Bachmann and D. Munz, "Unusual Potential Drop During Application of Electrical Potential Method in a Fracture Mechanics Test," *Journal of Testing and Evaluation* 4 (4), 257-260 (1976).

[51] S. Lee, T. Livne, and W. W. Gerberich, "The Acoustic-Emission Measurement of Cleavage Initiation Near The Ductile Brittle Transition-Temperature In Steel," *Scripta Metallurgica* 20 (8), 1137-1140 (1986).

[52] M. N. Pacey, M. N. James, and E. A. Patterson, "A new photoelastic model for studying fatigue crack closure," *Experimental Mechanics* 45 (1), 42-52 (2005).

[53] H. Toda, I. Sinclair, J. Y. Buffiere, E. Maire, K. H. Khor, P. Gregson, and T. Kobayashi, "A 3D measurement procedure for internal local crack driving forces via synchrotron X-ray microtomography," *Acta. Materialia* 52 (5), 1305-1317 (2004).

[54] B. Budiansky and J. W. Hutchinson, "Analysis of Closure in Fatigue Crack Growth," *Journal of Applied Mechanics-Transactions of the Asme* 45 (2), 267-276 (1978); Newman JC, "Mechanics of Fatigue Crack Growth. Special Technical Publication.," *Philadelphia ASTM* 590, 282-301 (1976).

[55] N. A. Fleck, "Finite-Element Analysis of Plasticity-Induced Crack Closure Under Plane-Strain Conditions," *Engineering Fracture Mechanics* 25 (4), 441-449 (1986); C. H. Wang, L. R. F. Rose, and J. C. Newman, "Closure of plane-strain cracks under large-scale yielding conditions," Fatigue & *Fracture of Engineering Materials & Structures* 25 (2), 127-139 (2002).

[56] R. C. McClung and H. Sehitoglu, "On the Finite-Element Analysis of Fatigue Crack Closure .1. Basic Modeling Issues," *Engineering Fracture Mechanics* 33 (2), 237-252 (1989).

[57] M. Andersson, C. Persson, and S. Melin, "Experimental and numerical investigation of crack closure measurements with electrical potential drop technique," *International Journal of Fatigue* 28 (9), 1059-1068 (2006).

[58] M. S. Ramos, M. V. Pereira, F. A. Darwish, S. H. Motta, and M. A. Carneiro, "Effect of single and multiple overloading on the residual fatigue life of a structural steel," Fatigue & Fracture of Engineering Materials & Structures 26 (2), 115-121 (2003); A. Steuwer, L. Edwards, S. Pratihar, S. Ganguly, M. Peel, M. E. Fitzpatrick, T. J. Marrow, P. J. Withers, I. Sinclair, K. D. Singh, N. Gao, T. Buslaps, and J. Y. Buffiere, "In situ analysis of cracks in structural materials using synchrotron X-ray tomography and diffraction," Nuclear Instruments & Methods in Physics Research Section B-Beam Interactions with Materials and Atoms 246 (1), 217-225 (2006); P. J. Withers, J. Bennett, Y. C. Hung, and M. Preuss, "Crack opening displacements during fatigue crack growth in Ti-SiC fibre metal matrix composites by X-ray tomography," *Materials Science and Technology* 22 (9), 1052-1058 (2006).

[59] Y. N. Sun, H. Choo, P. K. Liaw, Y. L. Lu, B. Yang, D. W. Brown, and M. A. M. Bourke, "Neutron diffraction studies on lattice strain evolution around a crack-tip during tensile loading and unloading cycles," Scripta Materialia 53 (8), 971-975 (2005); S. Y. Lee, R. I. Barabash, J. S. Chung, P. K. Liaw, H. Choo, Y. Sun, C. Fan, L. Li, D. W. Brown, and G. E. Ice, "Neutron and X-ray Microbeam Diffraction Studies around a Fatigue-Crack Tip after Overload," *Metallurgical and Materials*

Transactions a-Physical Metallurgy and Materials Science 39A (13), 3164-3169 (2008).

[60] S. Y. Lee, H. Choo, P. K. Liaw, E. C. Oliver, and A. M. Paradowska, "In situ neutron diffraction study of internal strain evolution around a crack tip under variable-amplitude fatigue-loading conditions," *Scripta Materialia* 60 (10), 866-869 (2009).

[61] S. Y. Lee, P. K. Liaw, H. Choo, and R. B. Rogge, "A study on fatigue crack growth behavior subjected to a single tensile overload Part I. An overload-induced transient crack growth micromechanism," Acta Materialia 59 (2), 485-494 (2011); S. Y. Lee, R. B. Rogge, H. Choo, and P. K. Liaw, "Neutron diffraction measurements of residual stresses around a crack tip developed under variable-amplitude fatigue loadings," Fatigue & Fracture of Engineering Materials & Structures 33 (12), 822-831 (2010); S. Y. Lee, Y. Sun, K. An, H. Choo, C. R. Hubbard, and P. K. Liaw, "Evolution of residual-strain distribution through an overload-induced retardation period during fatigue-crack growth," *Journal of Applied Physics* 107 (2) (2010).

[62] M. W. Barsoum, M. Radovic, T. Zhen, P. Finkel, and S. R. Kalidindi, "Dynamic Elastic Hysteretic Solids and Dislocations," *Physical Review Letters* 94 (8), 085501 (2005).

[63] Migliori, J. L. Sarrao, W. M. Visscher, T. M. Bell, M. Lei, Z. Fisk, and R. G. Leisure, "Resonant Ultrasound Spectroscopic Techniques for Measurement of the Elastic-Moduli Of Solids," *Physica B* 183 (1-2), 1-24 (1993).

[64] K. Zacharias, E. Balabanidou, I. Hatzokos, I. T. Rekanos, and A. Trochidis, "Microdamage evaluation in human trabecular bone based on nonlinear ultrasound vibro-modulation (NUVM)," J. Biomech. 42 (5), 581-586 (2009); M. A. Carpenter and Z. Y. Zhang, "Anelasticity maps for acoustic dissipation associated with phase transitions in minerals," *Geophys. J. Int.* 186 (1), 279-295 (2011).

[65] T. Adachi, Y. Kondo, A. Yamaji, S. H. Yang, and I. Y. Yang, "Nondestructive evaluation of micro-cracks in a ceramic ferrule by resonant ultrasound spectroscopy," Ndt&E Int 38 (7), 548-553 (2005).

[66] S. Agosta, R. G. Leisure, K. Foster, J. Markmann, and J. J. Adams, "Elastic moduli and internal friction of nanocrystalline Pd and PdSi as a function of temperature," *Philosophical Magazine* 88 (6), 949-958 (2008).

[67] M. R. Daymond, M. A. M. Bourke, R. B. VonDreele, B. Clausen, and T. Lorentzen, "Use of Rietveld refinement for elastic macrostrain

determination and for evaluation of plastic strain history from diffraction spectra," *J. Appl. Phys.* 82 (4), 1554-1562 (1997).

[68] Wen Huang, Rozaliya Barabash, Bjørn Clausen, and Peter Liaw, "Cyclic-Loading Induced Lattice-Strain Asymmetry in Loading and Transverse Directions," *Metallurgical and Materials Transactions* A 43 (5), 1454-1461 (2012).

[69] M. R. Daymond and P. J. Withers, "A new stroboscopic neutron diffraction method for monitoring materials subjected to cyclic loads: Thermal cycling of metal matrix composites," *Scr. Mater.* 35 (6), 717-720 (1996).

[70] Agnew S. R. Wu L., Brown D. W., Stoica G. M., Jain A., Clausen B., Fielden D. E., and Liaw P. K., "*Internal-stress relaxation and load redistribution during the twinning-detwinning-dominated cyclic deformation of a wrought magnesium alloy, ZK60A*," *Acta. Materialia* 56, 3699-3707 (2008).

[71] E-W. Huang, R. I. Barabash, W. Liu, Y.-L. Liu, J.-J. Kai, G. E. Ice, and P. K. Liaw, "Cyclic-loading-induced accumulation of geometrically necessary dislocations near grain boundaries in a nickel-based superalloy," *JOM* 61 (12), 51-56 (2009).

[72] Arya, G. K. Dey, V. K. Vasudevan, and S. Banerjee, "Effect of chromium addition on the ordering behaviour of Ni-Mo alloy: experimental results vs. electronic structure calculations," *Acta. Materialia* 50 (13), 3301-3315 (2002).

[73] Y. D. Wang, E. W. Huang, Y. Ren, Z. H. Nie, G. Wang, Y. D. Liu, J. N. Deng, H. Choo, P. K. Liaw, D. E. Brown, and L. Zuo, "In situ high-energy X-ray studies of magnetic-field-induced phase transition in a ferromagnetic shape memory Ni-Co-Mn-In alloy," *Acta. Materialia* 56 (4), 913-923 (2008).

[74] E. Wen Huang, "Evolution of microstructure in a nickel-based superalloy as a function of ageing time," *Philosophical magazine letters* 91 (7), 483 (2011).

[75] E. W. Huang, P. K. Liaw, L. Porcar, Y. Liu, Y. L. Liu, J. J. Kai, and W. R. Chen, "Study of nanoprecipitates in a nickel-based superalloy using small-angle neutron scattering and transmission electron microscopy," *Applied Physics Letters* 93 (16) (2008).

[76] R. I. Barabash, E. W. Huang, J. J. Wall, J. H. Wilkerson, Y. Ren, W. J. Liu, S. C. Vogel, G. E. Ice, L. M. Pike, and P. K. Liaw, "Texture crossover: Trace from multiple grains to a subgrain," *Materials Science*

and Engineering A-Structural Materials Properties Microstructure and Processing 528 (1), 3-10 (2010).

[77] C. and Von Dreele Larson, R. B., 2004.

[78] E. W. Huang, R. I. Barabash, Y. D. Wang, B. Clausen, L. Li, P. K. Liaw, G. E. Ice, Y. Ren, H. Choo, L. M. Pike, and D. L. Klarstrom, "Plastic behavior of a nickel-based alloy under monotonic-tension and low-cycle-fatigue loading," *International Journal of Plasticity* 24 (8), 1440-1456 (2008).

INDEX

A

advancement, 57
aging process, 61
amplitude, 20, 51, 52, 66, 71
annealing, 15, 17, 21, 23, 25
annihilation, 63
assessment, 64
asymmetry, 60, 63
atoms, 6, 13, 15, 16, 17, 19, 20, 21, 23, 26, 27, 28, 29, 30, 32, 33, 34, 35, 50, 61

B

banks, 59
base, 31, 32, 50, 64
beams, 37
behaviors, 40, 44, 47, 51, 53, 54
beneficial effect, 52
bone, 71
branching, 15, 52

C

carbon, 14, 15
ceramic, 71
challenges, 47, 50
chemical, 14, 17, 28, 29, 32
chemical bonds, 32
chromium, 72
closure, 52, 53, 54, 55, 64, 66, 67, 69, 70
community, 53
comparative analysis, 16
compliance, 52, 53, 54
composite material, 1, 2, 10, 37, 38, 41, 44, 45, 49, 65
composites, 70, 72
composition, 14, 15, 16, 20, 27, 32, 33, 34, 35
compounds, 2, 15, 27, 30, 32
compression, 44, 45
computer, 51
conductor, 44
cooling, 37, 38, 39
cooling process, 37, 38
coordination, 14, 30, 32
copper, 66
correlation, 15, 34
corrosion, 14
cracks, 51, 53, 55, 60, 70, 71
crystal structure, 14, 15, 20, 27, 34, 43
crystalline, 42
cycles, 54, 59, 60, 61, 70
cycling, 72

D

damages, 51, 57, 61
data processing, 18, 19

de Broglie wave rule, 3
decay, 15, 19, 23, 35
deformation, 27, 33, 34, 39, 52, 55, 57, 60, 61, 62, 64, 67, 72
degradation, 39, 51
delta, 69
depth, 2, 37
detection, 33, 66
detwinning, 49, 61, 65, 72
deuterium, 13, 14, 16, 20, 21, 23, 26, 27, 28, 30, 33, 34
diffraction, vii, 1, 2, 3, 4, 5, 6, 7, 8, 9, 10, 14, 15, 16, 17, 18, 20, 22, 23, 25, 26, 27, 28, 29, 32, 37, 38, 39, 42, 43, 44, 48, 49, 50, 53, 54, 55, 56, 57, 58, 59, 60, 61, 62, 63, 65, 70, 71, 72
direct measure, 57
dislocation, 48, 50, 51, 57, 62, 63, 65, 66, 69
disorder, 15, 26, 27, 34, 36
dispersion, 3, 35
displacement, 13, 21, 52, 54
distribution, 13, 14, 15, 21, 27, 34, 39, 40, 44, 51, 71
divergence, 27
domain structure, 23
drawing, 43

E

elastic deformation, 39, 40
elastic scattering, 2
electric current, 53
electromagnetic, 39, 44
electron(s), 48, 50, 54, 57
electron microscopy, 48, 57
electronic structure, 72
emission, 53
energy, 2, 16, 32, 49, 50, 62, 72
engineering, 3, 10, 44, 47, 48, 49, 50, 51, 55, 57, 61, 65
entropy, 66
environment(s), 3, 14, 59
environmental temperatures, 48
equilibrium, 21, 39, 40
evidence, 21, 23, 34
evolution, 48, 50, 57, 59, 60, 61, 62, 63, 70, 71
exposure, 19, 23

F

fabrication, 39, 43
fatigue, 47, 48, 49, 50, 51, 52, 53, 54, 55, 56, 57, 59, 60, 61, 62, 63, 64, 65, 66, 67, 68, 69, 70, 71, 73
ferromagnetic, 72
fixed detectors, 3
flight, 3, 6, 42
fluctuations, 35
force, 34, 39, 44, 48, 52, 53, 55
formation, 16, 24, 61
formula, 29, 51
France, 49
friction, 57, 71

G

geometry, 56
glasses, 66
grain boundaries, 72
gratings, 69
growth, 49, 50, 51, 52, 54, 64, 65, 66, 68, 70, 71
growth mechanism, 64
growth rate, 52

H

height, 30
history, 69, 72
homogeneity, 15, 26, 33
host, 23
human, 71
hydrides, 14, 30
hydrogen, 2, 13, 14, 15, 16, 17, 19, 20, 21, 26, 27, 30, 32, 34, 35
hydrogen atom(s), 13, 15, 16, 17, 19, 21, 26, 30, 32, 34

hysteresis, 38, 39, 40, 42, 45

I

ideal, 21, 23, 26, 44
identification, 2
impurities, 14
initiation, 51, 60, 66
insertion, 35
integrity, 49, 64
interaction effect(s), 52
interference, 2
interstitial atoms, 13, 15, 16, 27, 33, 34
inversion, 30
isotope, 14, 16, 27, 32
issues, 50, 53

L

lattice parameters, 8, 17, 23, 43
lead, 23, 37, 52
lifetime, 51, 52
light, 1, 2, 19, 31, 33, 48, 61
localization, 15
low temperatures, 27
lying, 21, 32

M

magnesium, 49, 61, 64, 65, 72
magnesium alloys, 61
magnet, 2, 39, 44
magnetic field, 37
magnetic materials, 2
magnetic structure, 2
mapping, 49, 55, 56, 57, 63
mass, 6
materials, vii, 1, 2, 3, 10, 36, 37, 38, 39, 43, 47, 48, 49, 51, 53, 57, 59, 65, 66, 70, 72
matrix, 38, 44, 49, 62, 65, 70, 72
matter, 20
measurement(s), 1, 2, 3, 4, 10, 37, 38, 39, 42, 44, 48, 49, 52, 54, 55, 56, 57, 58, 59, 60, 61, 62, 63, 69, 69, 70, 71
mechanical chopper, 3
mechanical properties, 38
melting, 14
melting temperature, 14
memory, 72
metals, 35, 48, 50, 67
methodology, 51
microscope, 54
microscopy, 62
microstructure, 48, 50, 51, 53, 54, 56, 57, 59, 62, 63, 72
microstructure measurements, 59
microstructures, 51, 62
modelling, 67
models, 39, 45, 50, 53, 62
modulus, 38, 44, 55
MTS, 59
multi-peak analysis, 1, 8, 9, 10, 38, 42, 43, 44
multiplication, 29

N

Nd, 54
neodymium, 2
Netherlands, 36
neutral, 48
neutron diffraction, vii, 1, 2, 3, 5, 6, 9, 10, 14, 15, 16, 17, 20, 23, 25, 26, 27, 29, 37, 39, 42, 43, 48, 49, 50, 53, 54, 55, 57, 60, 61, 62, 65, 71, 72
neutrons, 2, 35, 49, 55
nickel, 61, 64, 72, 73
nitrogen, 14, 15, 16, 17, 19, 20, 21, 23, 26, 27, 30, 34, 35
NRF, 63
nucleation, 64
nuclei, 16, 50
nucleus, 49

O

oscillation, 32
overlap, 2, 19

oxygen, 2, 14

P

parallel, 6, 30, 31, 32, 34, 56, 59
phase transformation, 35
phase transitions, 19, 71
photons, 50
physics, 1, 10
plastic deformation, 39, 49, 54, 55, 63, 64
plasticity, 48, 52, 53, 54, 64, 65, 67, 68
principles, 1, 10
probability, 51
propagation, 49, 51, 52, 53, 56, 64, 65, 66, 67, 68

Q

quartz, 17

R

radiation, 17, 48, 50, 63
radius, 28
reality, 51
recovery, 39, 40, 41
recovery process, 39
redistribution, 34, 40, 72
relaxation, 72
repulsion, 27
researchers, 53
resolution, 5, 23, 60, 69
response, 54, 61
retardation, 51, 52, 53, 54, 67, 71
room temperature, 4, 37, 38
roughness, 52
rules, 50
RUS, 57

S

safety, 47
scattering, 2, 6, 16, 18, 20, 26, 50, 56, 72
scattering intensity, 6
screw dislocations, 62
sensors, 66
severe stress, 52
shape, 49, 59, 62, 65, 72
shape-memory, 49, 65
showing, 56
signs, 13
silica, 17
simulation, 7, 48, 51
single crystals, 64
sintering, 17
SNS, 49, 58, 59
society, 48, 63
software, 59, 63
solid solutions, 13, 14, 15, 16, 17, 20, 23, 26, 27, 28, 29, 32, 33, 34, 35
solid state, 36
solution, 13, 14, 15, 16, 17, 19, 21, 23, 26, 27, 28, 29, 30, 32, 34, 35
specific gravity, 14
specifications, 59
spectroscopy, 57, 71
stability, 14, 27
state(s), 13, 19, 21, 26, 27, 39, 41, 42, 43, 51, 66
statistics, 42, 44
steel, 67, 70
storage, 2
stress, 8, 38, 39, 40, 41, 42, 44, 45, 49, 51, 52, 53, 54, 55, 56, 57, 64, 65, 68, 72
stress fields, 49, 52, 54, 65
structural characteristics, 13
structure, 2, 6, 8, 9, 14, 15, 16, 17, 18, 19, 21, 22, 23, 24, 26, 28, 29, 32, 33, 34, 38, 43, 50, 57, 59, 61, 62, 63
substitution, 20, 27
superconductivity, 2
superconductor, 39, 40, 41, 42, 44, 45
symmetry, 29, 61
synthesis, 6, 17, 36

T

techniques, 14, 53, 54, 57, 63, 64

technology, vii, 36
temperature, 2, 4, 13, 15, 16, 17, 19, 21, 22, 26, 27, 33, 34, 38, 39, 40, 41, 42, 45, 60, 61, 62, 71
tensile strength, 38
tension, 56, 73
testing, 60
texture, 8, 43, 45, 49, 57, 58, 59, 61, 62, 63
thermal expansion, 37, 38
thermal stability, 14
time frame, 59
time of flight (TOF), 3
titanium, 14, 15, 20, 28, 30, 35
transformation(s), 49, 61, 65
transmission electron microscopy, 72
twinning, 61, 72

U

ultrasound, 48, 57, 71
unique features, 50

V

vacancies, 34
vacuum, 42
variations, 44, 57
vector, 56

W

weak interaction, 50
websites, 56
wires, 37, 38, 43, 44

X

X-ray diffraction, 2, 3, 17, 21, 23, 24, 27, 54, 55

Y

yield, 59, 60, 67

Z

zirconium, 35